MW00760361

Origin of Temporal (t > 0) Universe

Origin of Temporal (t > 0) Universe

Connecting with Relativity, Entropy, Communication, and Quantum Mechanics

Francis T. S. Yu

CRC Press
Taylor & Francis Group
Boca Raton London New York

CRC Press is an imprint of the
Taylor & Francis Group, an **Informa** business

CRC Press
Taylor & Francis Group
6000 Broken Sound Parkway NW, Suite 300
Boca Raton, FL 33487-2742

© 2020 by Taylor & Francis Group, LLC

CRC Press is an imprint of Taylor & Francis Group, an Informa business

No claim to original U.S. Government works

Printed on acid-free paper

International Standard Book Number-13: 978-0-3674-1042-1 (Hardback)

This book contains information obtained from authentic and highly regarded sources. Reasonable efforts have been made to publish reliable data and information, but the author and publisher cannot assume responsibility for the validity of all materials or the consequences of their use. The authors and publishers have attempted to trace the copyright holders of all material reproduced in this publication and apologize to copyright holders if permission to publish in this form has not been obtained. If any copyright material has not been acknowledged please write and let us know so we may rectify in any future reprint.

Except as permitted under U.S. Copyright Law, no part of this book may be reprinted, reproduced, transmitted, or utilized in any form by any electronic, mechanical, or other means, now known or hereafter invented, including photocopying, microfilming, and recording, or in any information storage or retrieval system, without written permission from the publishers.

For permission to photocopy or use material electronically from this work, please access www.copyright.com (www.copyright.com/) or contact the Copyright Clearance Center, Inc. (CCC), 222 Rosewood Drive, Danvers, MA 01923, 978-750-8400. CCC is a not-for-profit organization that provides licenses and registration for a variety of users. For organizations that have been granted a photocopy license by the CCC, a separate system of payment has been arranged.

Trademark Notice: Product or corporate names may be trademarks or registered trademarks, and are used only for identification and explanation without intent to infringe.

Visit the Taylor & Francis Web site at
www.taylorandfrancis.com

and the CRC Press Web site at
www.crcpress.com

*All laws and paradoxes were made
to be broken or revised.*

Contents

Preface

In mathematics, every postulation needs to prove that there exists a solution before searching for the solution. Yet, in science, it seems that science does not have a criterion, as mathematics does, to prove first that a postulated science exists within our temporal universe. Without such a criterion, fictitious science emerges, as already have been happening in everyday event. And this is one the objectives for writing this book, in which I have shown that there exists a criterion for a postulated science whether or not it exists within our universe. I have started this book from Einstein's relativity to the creation of our temporal universe. In this book, I have shown that every subspace within our universe is created by energy and time, in which subspace and time are coexisted. The important aspect is that every science has to satisfy the boundary condition of our universe: causality and dimensionality. Following up with temporal universe, I have shown a profound relationship with second law of thermodynamics. Intriguing relationship between entropy and science and as well as communication with quantum limited subspace are parts of this book. Nevertheless, it must be the paradox of Schrödinger's Cat (which had debated by Einstein, Bohr, Schrödinger, and many others since 1935), which triggered me to discover that Schrödinger's quantum mechanics is a timeless machine, in which I have disproved that the fundamental principle of superposition within our universe. Since quantum mechanics is a virtual mathematics, I have shown that a temporal quantum machine can, in principle, be built on the top of a temporal platform.

Nevertheless, the essence of this book is to show everything has a price within our temporal ($t > 0$) universe: energy and time. And science exists for new discoveries and renovation. For instance, one of the world greatest astrophysicists told us that Black Hole provides an invisible channel connecting to other universe, which is absurd. Or there exists a theory of all theories, would you take it seriously? A mathematician found a ten-dimensional subspace within our universe, won't you curiously enough to find out whether it exists within our universe or not? After all, we are all humans; imperfect and limited.

This is all about science; all laws and paradoxes of science were made to be broken or revised.

In short, I would like to express my appreciation to colleagues and friends who have viewed my articles and for their valuable comments; without their encouragement, this book would not have been written.

About the Author

Francis T. S. Yu received his B.S.E.E. degree from Mapua Institute of Technology Manila, Philippines, and his M.S. and Ph.D. degrees in Electrical Engineering from the University of Michigan. He has been a consultant to several industrial and governmental laboratories. He is an active researcher in the fields of optical signal processing, holography, information optics, optical computing, neural networks, photorefractive optics, fiber sensors, and photonic devices. He has published over 500 papers, in which over 300 are referred. He has served as an associate editor, editorial board member, and a guest editor for various international journals. He is the author and coauthor of nine books. Dr. Yu is a life fellow of IEEE and a fellow of OSA, SPIE, and PSC.

chapter one

From Relativity to Discovery of Temporal (t > 0) Universe

1.1 Introduction

One of the most intriguing variables in science must be time. Without time, there would be no physical substances, no space, **and** no life. In other words, time and substance have to be coexisted. In this chapter, I will start with Einstein's relativity theory to show his famous energy equation, in which we will show that energy and mass can be traded. Since mass is equivalent to energy and energy is equivalent to mass, we see that mass can be treated as an energy reservoir. We will show that any physical space cannot be embedded in an absolute empty space and it cannot have any absolute empty subspace in it and empty space is a timeless (i.e., t = 0) space. We will show that every physical space has to be fully packed with substances (i.e., energy and mass) and we will show that our universe is a subspace within a more complex space. In which we see that, our universe could have been one of the many universes outside our universal boundary. We will also show that it takes time to create a subspace, and it cannot bring back the time that has been used for the creation. Since all physical substances exist with time, all subspaces are created by time and substances (i.e., energy and mass). This means that our cosmos was created by time with a gigantic energy explosion, for which every subspace is coexisted with time. This means that, without time, the creation of substances would not have happened. Of which we see that, our universe is in a temporal (i.e., t > 0) space, and it is still expanding based on current observation. This shows that our universe has not reached its half-life yet, as we have accepted the big bang creation. We are not alone with almost absolute certainty. Someday, we may find a planet that once upon a time had harbored a civilization for a period of light years. In short, the burden of a scientific postulation is to prove that there exists a solution within out temporal universe; otherwise it is not real or virtual as mathematics does.

Professor Hawking was a world-renowned astrophysicist, a respected cosmic scientist, and a genius who passed away last year on March 14, 2018. As you will see that our creation of universe was started with the same root of Big Bang explosion, but it is not a sub-universe of Hawking's. You may see from this chapter the creation of temporal universe is somewhat different from Hawking's creation. One of the examples is that our temporal universe creation was started within a non-empty subspace, instead of an empty space as Hawking did.

1.2   Relativity to Einstein Energy Equation

The essence of Einstein's Special Theory of Relativity [1] is that time is a relative quantity with respect to velocity as given by,

$$\Delta t' = \frac{\Delta t}{\sqrt{1 - v^2/c^2}} \tag{1.1}$$

where $\Delta t'$ is the relativistic time window as compared with a standstill subspace, Δt is the time window of the standstill subspace, v is the velocity of a moving subspace, and c is the velocity of light.

In which, we see that the time window $\Delta t'$ of a moving subspace, with respect to the time window Δt of a standstill subspace, appears to be wider as velocity of the moving subspace increases. In other words, velocity of a moving subspace changes the relative time speed as with respect to a standstill subspace. For instance, the time-speed goes slower for a moving subspace as with respect to a standstill subspace. Of which we see that time-speed within the subspaces is invariant or constant. In other words, the speed of time goes as it is within the subspaces, but relatively different between the subspaces at different velocities. As a matter of fact, the speed of time within a subspace is governed by the speed of light (such as 1 second, 2 second, ...) as will be seen how our temporal universe was created.

Equivalently, Einstein relativity equation can be shown in terms of relative mass as given by,

$$m = \frac{m_0}{\sqrt{(1 - v^2/c^2)}} = m_0(1 - v^2/c^2)^{-1/2} \tag{1.2}$$

where m is the effective mass (or mass in motion) of a particle, m_0 is the rest mass of the particle, v is the velocity of the moving particle, and c is the speed of light. In other words, the effective mass (or mass in motion) of a particle increases at the same amount with respect to the relative time window increases.

With reference to the binomial expansion, Eq. (1.2) can be written as,

$$m = m_0(1 + \frac{1}{2} \cdot \frac{v^2}{c^2} + \textit{terms of order } \frac{v^4}{c^4}) \tag{1.3}$$

By multiplying the preceding equation with the velocity of light c^2 and noting that the terms with the orders of v^4/c^2 are negligibly small, above equation can be approximated by,

$$m \approx m_0 + \frac{1}{2}m_0 v^2 \frac{1}{c^2} \tag{1.4}$$

which can be written as,

$$(m - m_0)c^2 \approx \frac{1}{2}m_0 v^2 \tag{1.5}$$

The significant of the preceding equation is that m-m_0 represents an increase in mass due to motion, which is the kinetic energy of the rest mass m_0. And $(m$-$m_0)c^2$ is the extra energy gain due to motion.

What Einstein postulated, as I remembered, is that there must be energy associated with the mass even at rest. And this was exactly what he had proposed,

$$\varepsilon \approx mc^2 \tag{1.6}$$

where ε represents the total energy of the mass and

$$\varepsilon_0 \approx m_0 c^2 \tag{1.7}$$

the energy of the mass at rest, where $v = 0$ and $m \approx m_0$.

In which we see that Eq. (1.6) or equivalently Eq. (1.7) is the well-known Einstein Energy Equation.

1.3   Time and Energy

One of the most enigmatic variables in the laws of science must be "time." So what is time? Time is a variable and not a substance. It has no mass, no weight, no coordinate, no origin, and it cannot be detected or even be seen. Yet time is an everlasting existed variable within our known universe. Without time, there would be no physical matter, no physical space, and no life. The fact is that every physical matter is associated with time, including our universe. Therefore, when one is dealing with science, time is one of the most enigmatic variables that ever presence and cannot be simply ignored. Strictly speaking, all the laws of science as well every physical substance cannot be existed without the existence of time.

On the other hand, energy is a physical quantity that governs every existence of substance, which includes the entire universe. In other words without the existence of energy, there would be no substance and no universe! Nonetheless, based on our current laws of science, all the substances were created by energy and every substance can also be converted back to energy. Thus energy and substance are exchangeable, but it requires some physical conditions (e.g., nuclei and chemical interactions and others) to make the conversion started. Since energy can be derived from mass, mass is equivalent to energy. Hence, every mass can be treated as an **energy reservoir**. The fact is that, our universe is compactly filled with mass and energy. Without the existence of time, the trading (or conversion) between mass and energy **could not have happened.**

1.4   Time-Dependent Energy Equation

Let us now start with the Einstein's Energy Equation, which was derived by his Special Theory of Relativity [1] as given by,

$$\varepsilon \approx mc^2 \tag{1.8}$$

where m is the rest mass and c is the velocity of light.

Since all the laws in science are approximations, we have intentionally used an approximated sign. Strictly speaking, the energy equation should be more appropriately presented with an inequality sign as described by,

$$\varepsilon < mc^2 \tag{1.9}$$

This means that, in practice, the total energy should be smaller or at most approaching to the rest mass m times square of light speed (i.e., c^2).

In view of the Einstein's Energy Equation of Eq. (1.8), we see that it is a singularity-point approximated and timeless equation (i.e., t = 0). In other words, the equation needs to convert into a temporal (i.e., t > 0) representation or time-dependent equation for the conversion to take place from mass into energy. For which we see that, without the inclusion of time variable, the conversion would not have taken place. Nonetheless, Einstein's Energy Equation represents the **total amount of energy** that can be converted from a rest mass m, in which every mass can be viewed as an **energy reservoir**. Thus by incorporating the time variable, the Einstein's Energy Equation can be represented by a partial differential equation as given by [2],

$$\frac{\partial \varepsilon(t)}{\partial t} = c^2 \frac{\partial m(t)}{\partial t}, \qquad t > 0 \tag{1.10}$$

where $\partial\varepsilon(t)/\partial t$ is the rate of increasing energy conversion, $\partial m(t)/\partial t$ is the corresponding rate of mass reduction, c is the speed of light, and t > 0 represents a forward time variable. In which we see that a time-dependent equation exists at time t > 0 and represents a forwarded time variable that only occurs after time excitation at t = 0. Incidentally, this is the well-known causality constraint (i.e., t > 0) [3] as imposed by our universe.

1.5 Trading Mass and Energy

One of the important aspects in the preceding Eq. (1.10) must be that energy and mass can be **traded**, for which the **rate of energy conversion from a mass** can be written in terms of Electro-Magnetic (EM) radiation or Radian Energy as given by [4],

$$\frac{\partial\varepsilon}{\partial t} = c^2\frac{\partial m}{\partial t} = [\nabla \cdot S(v)] = -\frac{\partial}{\partial t}[\frac{1}{2}\in E^2(v) + \frac{1}{2}\mu H^2(v)], \quad t>0 \qquad (1.11)$$

where $\in$ and μ are the permittivity and the permeability of the physical space, respectively, v is the radian frequency variable, $E^2(v)$ and $H^2(v)$ are the respective electric and magnetic field intensities, the **negative sign** represents the **out-flow** energy per unit time from an unit volume, $(\nabla\cdot)$ is the divergent operator, and S is known as the **Poynting Vector** or **Energy Vector** of an electro-magnetic radiator [4] as can be shown by $S(v) = E(v) \times H(v)$. Again we note that it is a time-dependent equation with t > 0 added to present the causality constraint. In view of the preceding equation, we see that radian energy (i.e., radiation) diverges from the mass, as mass reduces with time. In other words, we see that Eq. (1.11) is not just a piece of mathematical formula; it is a **symbolic** representation, a **description**, a **language**, a **picture**, or even a **video**, as it can be seen that it has transformed from a point-singularity approximation into three-dimensional representation and it is continually expanding as time moves on.

One significant implication of Eq. (1.11) is that trading from mass to energy is occurring within a non-empty physical subspace of $\in$ and μ (i.e., permittivity and permeability); otherwise the velocity of the expanding energy density will be unlimited (i.e., infinitely large). In other words, the space of the mass to energy creation has to be non-empty; otherwise electro-magnetic waves (i.e., radiant energy) simply cannot propagate within a space without medium (i.e., empty). We stress that practically all the theories concerning the creation of our universe were started within an absolute empty space. But we know that empty and non-empty spaces cannot coexist, in which we see that our universe has to be created within a non-empty subspace, instead of using an empty space as normally assumed.

Similarly, the **conversion from energy to mass** can also be presented as,

$$\frac{\partial m}{\partial t} = \frac{1}{c^2}\frac{\partial \varepsilon}{\partial t} = -\frac{1}{c^2}[\nabla \cdot S(v)] = \frac{1}{c^2}\frac{\partial}{\partial t}[\frac{1}{2} \in E^2(v) + \frac{1}{2}\mu H^2(v)] \quad , \quad t > 0$$

(1.12)

The major difference of this equation, as compared with former Eq. (1.11), must be the energy convergent operator $-\nabla \cdot S(v)$, where we see that the rate of energy as in the form of EM radiation **converges** into a small volume for the mass creation, instead of diverging from the mass. Since mass creation is inversely proportional to c^2, it requires a huge amount of energy to produce a small quantity of mass. Nevertheless, in view of the cosmological environment, availability of huge amount of energy has never been a problem.

Incidentally, **Black Hole** [5, 6] can be considered as one of the **energy convergent operators**. Instead the convergent force is more relied on the Black Hole's intense gravitational field. Black Hole is still remaining an intriguing physical substance to be known. Its gravitational field is so intense even light cannot be escaped.

By the constraints of current laws of science, the observation is limited by the speed of light. If light is totally absorbed by the black hole, it is by **no mean that black hole is an infinite energy sink** [6]. Nonetheless, every black hole can actually be treated as an energy convergent operator, which is responsible for the eventuality in part of energy to mass conversion, where an answer is remained to be found.

1.6   *Physical Substances and Subspaces*

In our physical world, every matter is a substance that includes all the elemental particles, electric, magnetic, gravitation fields, and energy. The reason is that they were all created by means of energy or mass. Our physical space (e.g., our universe) is fully compacted with substances (i.e., mass and energy) and left no **absolute empty subspace** within it. As a matter of fact, all physical substances exist with time and no physical substance can exist forever or without time, which includes our universe. Thus, without time, there would be no substance and no universe. Since every physical substance described itself as a **physical space**, it is constantly changing with respect to time. The fact is that every physical substance is itself a **temporal space** (or a **physical subspace**), as this will be discussed in the subsequent sections.

In view of physical reality, every physical substance cannot be existed without time; thus, if there is no time, all the substances that include all the building blocks in our universe and the universe itself **cannot be existed**. On the other hand, time cannot be existed without the existence of substance or substances. Therefore, time and substance must be **mutually coexisted** or **inclusively existed**. In other words, substance and time have to be **simultaneously existed** (i.e., one cannot be existed without the other). Nonetheless, if our universe has to be existed with time, then our universe will eventually **get old and die**. So the **aspects of time** would not be as simple as we have known. For example, for the species living in a far distant galaxy moves closer to the speed of light, their time goes somewhat slower relatively to ours [1]. Thus, we see that the relativistic aspects of time may not be the same at different subspaces in our universe (e.g., at the edge of our universe).

Since substances (i.e., mass) were created by energy, energy and time have to be simultaneously existed. As we know that every conversion, either from mass to energy or from energy to mass, cannot get started without the inclusion of time. Therefore, time and substance (i.e., energy and mass) have to be simultaneously existed. Thus we see that, all the physical substances, including our universe and us, are coexisted with time (or function of time); and time and subspace (i.e., substance) are mutually dependent. In other words; time is a subspace dependent variable and subspace is a time dependent vaiable.

1.7 Absolute Empty and Physical Subspaces

Let us define various subspaces in the following, as they will be used in the subsequent sections.

An absolute empty space has no time, no substance, no coordinate, not event bounded or unbounded. It is a virtual space timeless space (i.e., $t = 0$) and it does not exist in practice.

A physical space is a space described by dimensional coordinates, existed in practice, compactly filled with substances, supported by the current laws of science and the rule of time (i.e., time is a dependent forward variable and cannot move backward; $t > 0$). In which we see that, physical space and absolute empty space are **mutually exclusive**: Where any physical space cannot be embedded in an absolute empty space and it cannot have any absolute empty subspace in it. In other words, physical space is a temporal ($t > 0$) space in which time is a dependent forward variable (i.e., $t > 0$), while absolute empty space is a timeless space (i.e., $t = 0$) in which nothing is in it.

A temporal space is a time variable physical space supported by laws of science and rule of time (i.e., $t > 0$). In fact, **all physical spaces are temporal spaces (i.e., $t > 0$)**.

A spatial space is a space described by dimensional coordinates and may not be supported by laws of science and the rule of time (e.g., a mathematical virtual space).

A virtual space is an imaginary space and it is generally not supported by the laws of science and the rule of time, in which only mathematicians can make it happen.

As we have noted, absolutely empty space cannot be existed in physical reality, as every physical space needs to be completely filled with substances and left no absolutely empty subspace within it, for which **every physical space is created by substances**. For example, our universe is a gigantic physical space created by mass and energy (i.e., substances) and has no empty subspaces in it. Yet, in physical reality all the masses (and energy) are existed with time. Without the existence of time, then there would be no mass, no energy, and no universe. Thus, we see that every physical substance is **coexisted** with time. As a matter of fact, every physical subspace is a **Temporal Subspace (i.e., t > 0)**, which includes us and our universe.

One of the most convincing examples in describing a temporal subspace is looking at clock ticks: We see that as the clock ticks, the whole phyical system (i.e.,subspace) of the clock is also changing (i.e., ticking) with time. For which we see that every passages of tick cannot bring back the precise system of the clock before the tick. In other words, as we age we cannot change back our physical being (i.e., our subspace) before the second that has aged us. Since time is coexisted with every subspace (i.e., substance) which is including us, we see that time is real and it is "not" an illusion.

Since a physical space cannot be embedded within an absolute empty space and it cannot have any absolute empty subspace in it [7], our universe must be embedded in a **more complex physical space**. If we accepted our universe is embedded in a more complex space, then our universe must be a **bounded subspace**.

How about time? Since our universe is embedded in a more complex space, the complex space may share the same rule of time (i.e., t > 0). However, the complex-space that embeds our universe may not have the same laws of science as ours, but may have the same rule of time (i.e., t > 0); otherwise our universe would not be bounded. Nevertheless, whether our universe is bounded or not bounded is not the major issue of our current interest, since it takes a deeper understanding of our current universe before we can move on to the next level of complex space revelation. It is however our aim, abiding within our current laws of science, to investigate the essence of time as the **enigma origin of our universe**.

1.8   Time and Physical Space

One intriguing question in our life must be the existence of time. So far, we know that time comes from nowhere, and it can only move forward, not backward, not even stand still (i.e., t = 0). Although time may relatively slow down somewhat, based on Einstein's special theory of relativity [1], so far time cannot move backward, cannot even stand still [i.e., a timeless (t = 0) space]. As a matter of fact, time is moving at a constant rate within our subspace and it cannot move faster or slower. We stress that time moves at the same rate within any subspace within the universe even closer the boundary of our universe, but the difference is the relativistic time. Since time is ever existed, then how do we know there is a physical space? One answer is that there is a profound connection between time and physical space. In other words, if there is no time, then there would be no physical space. A physical space is in fact a temporal (i.e., t > 0) space, in contrast to a virtual space. Temporal space can be described by time, while virtual space is an imaginary space without the constraint of time. In which, temporal space is supported by the laws of science, whereas virtual space is not.

A television video image is a typical example of trading time for space. For instance, each TV displayed image of (dx, dy) takes an amount of time to be displayed. Since time is a forward-moving variable, it cannot be traded back at the expense of a displayed image (dx, dy). In other words, it is time that determines the physical space, and it is not the physical space that can bring back the time that has been expended. And it is the size (or dimension) of space that determines the amount of time required to create the space (dx, dy). In which, time is distance and distance is time within a temporal space. Based on our current constraints of science, the speed of light is the limit. Since every physical space is created by substances, a physical space must be described by the speed of light. In other words, the dimension of a physical space is determined by the velocity of light, where the space is filled with substances (i.e., mass and energy). And this is also the reason that speed of time (e.g., 1 second, 2 second, …) is determined by the speed of light.

Another issue is why the speed of light is limited. It is limited because our universe is a gigantic physical space that is filled with substances that cause a time delay on an EM wave's propagation. Nevertheless, if there were physical substances travel beyond the speed of light (which remains to be found), their velocities would also be limited, since our physical space is fully compacted with physical substances and it is a temporal (i.e., t > 0) space. Let me further note that a substance can travel in space without a time delay if and only if the space is absolutely empty (i.e., timeless; t = 0), since distance is time (i.e., d = ct, t = 0). However, absolute empty space cannot exist in practice, since every physical space (including our universe) has to be fully filled with substances (i.e., energy and mass), with no empty subspace left within it. Since every physical subspace is temporal (i.e., t > 0), we see that timeless and temporal spaces are mutually exclusive.

1.9   Electro-Magnetic and Laws of Physics

Strictly speaking, all our laws of physics are evolved within the regime of EM science. Besides, all physical substances are part of EM-based science, and all the living species on the Earth are primarily dependent on the source of energy provided by the Sun. About 78% of the sunlight that reaches the surface of our planet is well concentrated within a narrow band of visible spectrum. In response to our species' existence, which includes all living species on Earth, a pair of visible eyes (i.e., antennas) evolved in humans, which help us for our survival. And this narrow band of visible light led us to the discovery of an even wider band of EM spectral distribution in nature. It is also the major impetus allowing us to discover all the physical substances that are part of EM-based physics. In principle, all physical substances can be observed or detected with EM interaction, and the speed of light is the current limit.

Then there is question to be asked, why the speed of light is limited. A simple answer is that our universe is filled with substances that limit the speed of light. The energy velocity of an electromagnetic wave is given by [4],

$$v = \frac{1}{\sqrt{\mu\varepsilon}} \tag{1.13}$$

where (μ, ε) are the permeability and the permittivity of the medium. We see that the velocity of light is shown by,

$$c = \frac{1}{\sqrt{\mu_0\varepsilon_0}} \tag{1.14}$$

where (μ_0, ε_0) are the permeability and the permittivity of the space.

In view of Eq. (1.13), it is apparent that the velocity of electromagnetic wave (i.e., speed of light) within an empty subspace [i.e., timeless (t = 0) space] is instant (or infinitely large) since distance is time (i.e., d = c·t; t = 0).

A picture that is worth more than a thousand words [7] is a trivial example to show that EM observation is one of the most efficient aspects in information transmission. Yet, the ultimate physical limitation is also imposed by limitation of the EM regime, unless new laws of science emerge. The essence of Einstein's energy equation shows that mass and energy are exchangeable. It shows that energy and mass are equivalent, and energy is a form of EM radiation in view of Einstein's equation. We further note that all physical substances within our universe were created from energy and mass, which includes the dark energies [8] and dark matter [9]. Although the dark substances may not be observed directly using EM interaction, we may indirectly detect their existence, since they are basically energy-based substances (i.e., EM-based science). It may be interesting to note that our current universe is composed of 72% dark energy, 23% dark matter, and 5% other physical substances. Although dark matter contributes about 23% of our universe, it represents a total of 23% of gravitational fields. With reference to Einstein's energy equation Eq. (1.8), dark energy and dark matter dominate the entire universal energy reservation, well over 95%. Furthermore, if we accept the Big Bang theory for our universe creation [10], then creation could have been started with Einstein's time-dependent energy formula of Eq. (1.11) as given by,

$$\frac{\partial \varepsilon}{\partial t} = c^2 \frac{\partial M}{\partial t} = [\nabla \cdot S(v)] = -\frac{\partial}{\partial t} [\frac{1}{2} \in E^2(v) + \frac{1}{2} \mu H^2(v)] , \quad t > 0 \qquad (1.15)$$

where $[\nabla \cdot S(v)]$ represents a divergent energy operation. In this equation, we see that a broad-spectral-band intense radian energy diverges (i.e., explodes) at the speed of light from a compacted matter M, where M represents a gigantic mass of energy reservoir. It is apparent that the creation is ignited by time and the exploded debris (i.e., matter and energy) starts to spread out in all directions, similar to an expanding air balloon. The boundary (i.e., radius of the sphere) of the universe expands at the speed of light, as the created debris is disbursed. It took about 15 billion chaotic light years [11, 12] to come up with the present state of constellation, in which the boundary is still expanding at the speed of light beyond the current observation. With reference to a recent report using the Hubble Space Telescope, we can see galaxies about 14 billion light years away from us. This means that the creation process is not stopping yet, and at the same time the universe might have started to de-create itself, since the Big Bang started, due to intense convergent gravitational forces from all the newly created debris of matter (e.g., galaxies and dark matter). To wrap up this section, we would stress that one of the viable aspects of Eq. (1.15) is the transformation from a spatially dimensionless equation to a space–time function (i.e., ∇·S); it describes how our universe was created with a huge explosion. Furthermore, the essence of Eq. (1.15) is not just a piece of mathematical formula; it is a symbolic representation, a description, a language, a picture, or even a video as may be seen from its presentation, in which we can visualize how our universe was created: from the theory of relativity to Einstein's energy equation and then to a temporal (t > 0) space creation.

Another important aspect of Eq. (1.15) I would like to stress once and again is that the creation is not started from an absolute empty space, since the speed of radiant energy density (i.e., the last term in this equation) is limited by the speed of light. As in contrast with normal, assume that the Big Bang creation was started within an empty space. Since empty and non-empty spaces are mutually exclusive, then the Big Bang explosion has to be started from a non-empty space as we can see from this equation. I would further note that only mathematicians and theoretical physicists can virtually implant a physical explosion within an empty space, since theoretical physics is a kind of mathematics. Even though we allow that a virtual Big Bang explosion can happen within an empty space, then the propagation of the explosive energy density (i.e., the electro-magnetic wave) will be instant or unlimited, for which our universe will not be a bounded sub-universe. And this is one of several differences of the temporal universe creation as compared with all the other theories of universal creations, which was created from an empty space.

1.10   *Trading Time and Subspace*

Let us now take one of the simplest connections between physical subspace and time [13]:

$$d = vt \tag{1.16}$$

where d is the distance, v is the velocity and t is the time variable. Notice that this equation may be one of the most profound connections between time and physical space (or temporal space). Therefore, a three-dimensional (Euclidean) physical (or temporal) subspace can be described by

$$(dx, dy, dz) = (vx, vy, vz)t \tag{1.17}$$

where (vx, vy, vz) are the velocity vectors and t is the time variable. Under the current laws of science, the speed of light is the limit. Then, by replacing the velocity vectors equal to the speed of light c, a temporal space can be written as

$$(dx, dy, dz) = (ct, ct, ct) \tag{1.18}$$

Thus, we see that time can be traded for space and space cannot be traded for time, since time is a forward variable (i.e., t > 0). In other words, once a section of time Δt is expended, we cannot get it back. Needless to say, a spherical temporal space can be described by

$$r = ct \qquad (1.19)$$

where radius r increases at the speed of light c. Thus, we see that the boundary (i.e., edge) of our universe is determined by radius r, which is limited by the light speed, as illustrated in a composite temporal-space diagram of Figure 1.1. In view of this figure, we see that our universe is expanding at the speed of light well beyond the current observable galaxies. Figure 1.2 shows a discrete temporal-space diagram, in which we see that the size of our universe is continuously expanding as time moves forward (i.e., t > 0). Assuming that we have already accepted the Big Bang creation, sometime in the future (i.e., billions of light years later) our universe will eventually stop expanding and then start to shrink back, preparing for the next cycle of Big Bang explosion. The forces for the collapsing universe are mainly due to the intense gravitational field, mostly from giant black holes and matter that were derived from merging (or swallowing) with smaller black holes and other debris (i.e., physical substances). Since a black hole's gravitational field is so intense, even light cannot be escaped; however, a black hole is by no means an infinite energy reservoir. Eventually, the storage capacity of a black hole will reach a limit for explosion, as started for the mass-to-energy and debris creation.

In other words, there will be one dominant giant black hole within the shrinking universe to initiate the next cycle of universe creation. Therefore, every black hole can be treated as a convergent energy sink, which relies on its intense gravitation field to collect all the debris of matter and energies. Referring to Big Bang creation, a gigantic energy explosion was the major reason for the universe's creation. In fact, it can be easily discerned that the creating process has never slowed down since the birth of our universe, as we see that our universe is still continuingly expanding even today. This is by no means an indication that all the debris created came from the Big Bang's energy (e.g., mc^2); there might have been some leftover debris from a preceding universe. Therefore, the overall energy within our universe cannot be restricted to just the amount that came from the Big Bang creation. In fact, the conversion processes between mass and energy have never been totally absent since the birth of our universe, but they are on a much smaller scale. In fact, right after birth, our universe started to slow down the divergent process due to the gravitational forces produced by the created matter. In other words, the

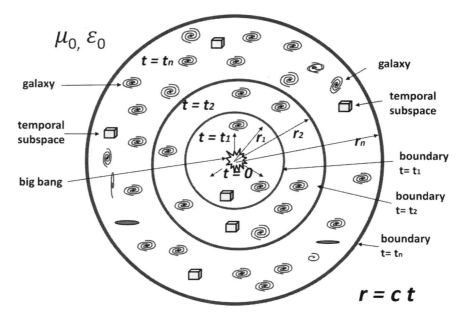

Figure 1.1   Composite temporal space universe diagrams. r = ct, r is the radius of our universe, t is time, c is the velocity of light, ε_0 and μ_0 are the permittivity and permeability of the space.

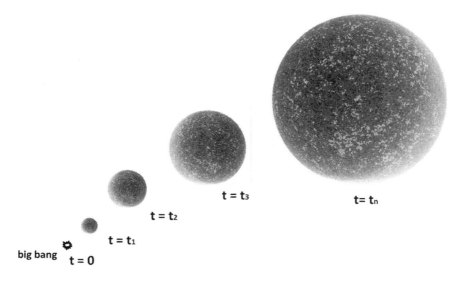

Figure 1.2   Discrete temporal universe diagrams; t is time.

universe will eventually reach a point when overall divergent forces will be weaker than the convergent forces, which are mostly due to gravitational fields coming from the newly created matter, including black holes. As we had mentioned earlier, our universe currently has about 23% dark matter, which represents about 23% of the gravitational fields within the current universe. The intense localized gravitational field could have been produced from a group or a giant black hole, derived from merging with (or swallowing up) some smaller black holes, nearby dark matter and debris. Since a giant black hole is not an infinite energy sink, eventually it will explode for the next cycle of universal creation. And it is almost certain that the next Big Bang creation will not occur at the same center of our present universe. One can easily discern that our universe will never shrink to a few inches in size, as commonly speculated. It will, however, shrink to a smaller size until one of the giant black holes (e.g., swallowed-up sufficient physical debris) reaches the Big Bang explosive condition to release its gigantic energy for the next cycle of universal creation. The speculation of a possible collapsing universe remains to be observed. Nonetheless, we have found that our universe is still expanding, as observed by the Doppler shifts of the distant galaxies at the edge of our universe, about 15 billion light years away [11,12]. This tells us that our universe has not reached its half-life yet. In fact, the expansion has never stopped since the birth of our universe, and our universe has also been started to de-create since the Big Bang started, which is primarily due to convergent gravitational forces from the newly created debris (e.g., galaxies, black holes and dark matter).

1.11   Relativistic Time and Temporal (t > 0) Space

Relativistic time at a different subspace within a vast universal space may not be the same as that based on Einstein's special theory of relativity [1]. Let us start with the relativistic time dilation as given by,

$$\Delta t' = \frac{\Delta t}{\sqrt{1 - v^2/c^2}} \tag{1.20}$$

where $\Delta t'$ is the relativistic time window, compared with a standstill subspace; Δt is the time window of a standstill subspace; v is the velocity of a moving subspace; and c is the velocity of light. We see that time dilation $\Delta t'$ of the moving subspace, relative to the time window of the standstill subspace Δt, appears to be wider as velocity increases. For example, a 1-second time window Δt is equivalent to 10-second relative time window $\Delta t'$. The means that 1-second time expenditure within the moving subspace is relative to about 10-second time expenditure within the standstill subspace. Therefore, for the species living in an environment that travels closer to the speed of light (e.g., at the edge of the universe), their time appears to be slower than ours, as illustrated in Figure 1.3. In this figure, we see an old man traveling at a speed closer to the velocity of light; his relative observation time window appears to be wider as he is looking at us, and the laws of science within his subspace may not be the same as ours.

Two of the most important pillars in modern physics must be the Einstein's Relativity Theory and the Schrödinger's Quantum Mechanics [13]. One is dealing with very large objects (e.g., universe) and the other is dealing with very small particles (e.g., atoms). Yet, there exists a profound connection between them by means of the Heisenberg's Uncertainty Principle [14]. In view of the Uncertainty Relation, we see that every temporal subspace takes a section of time Δt and an amount of energy ΔE to create. Since we cannot create something from nothing, everything needs an amount of energy ΔE and a section of time Δt to make it happen. By referring to the Heisenberg Uncertainty Relation, it is given as,

$$\Delta E \cdot \Delta t \geq h \tag{1.21}$$

where h is the Planck's constant. We see that every subspace is limited by ΔE and Δt. In other words, it is the h region, but not the shape, that determines the boundary condition. For example, the shape can be either elongated or compressed, as long as it is larger than h region.

Incidentally, the uncertainty relationship of Eq. (1.21) is also the limit of reliable bit information transmission as pointed out by Dennis Gabor in 1950 [15]. Nonetheless, the connection with the Special Theory

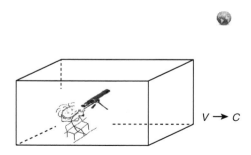

Figure 1.3   Effects on Relativistic Time.

of Relativity is that the creation of a subspace near the edge of our universe will take a short relative time with respect to our planet earth, since $\Delta t' > \Delta t$. The "relativistic" uncertainty relationship within the moving subspace, as with respect to a standstill subspace, can be shown as,

$$\Delta E \cdot \Delta t' \left[1 - (v/c)^2\right]^{1/2} \geq h \tag{1.22}$$

where we see ΔE energy is conserved. Thus a narrower time-width can be achieved as with respect to standstill subspace. This is precisely possible that one can exploit for time-domain digital communication, as from ground station to satellite information transmission.

One the other hand, as from satellite to ground station information transmission, we might want to use digital-bandwidth (i.e., Δv) instead. This is a frequency-domain information transmission strategy, as in contrast with time-domain, which has not been exploited yet. In which the "relativistic" uncertainty relationship within the standstill subspace as with respect to the moving subspace can be written as,

$$\frac{\Delta E \cdot \Delta t}{\sqrt{1 - \left(\frac{v}{c}\right)^2}} \geq h \tag{1.23}$$

Or equivalently we have,
$$\frac{\Delta v \cdot \Delta t}{\sqrt{1 - \left(\frac{v}{c}\right)^2}} \geq 1 \tag{1.24}$$

In which we see that a narrower bandwidth Δv can be in principle used for frequency domain communication.

1.12 Time and Physical Space

Every physical (or temporal) subspace is created by substances (i.e., energy and mass) and substances coexist with time. In this context, we see that our universe was essentially created by time and energy, and the universe is continuously evolving (i.e., changing) with time. Although relativistic time may not be the same at the different subspaces within our universe, the rule of time may remain the same. As for the species living closer to the speed of light, relativistic time may not be noticeable to them, but their laws of science within their subspace may be different from ours. Nonetheless, our universe was simultaneously created by time and a gigantic energy explosion. Since our universe cannot be embedded in an empty space, it must be embedded in a more complex space that remains to be found. From an inclusive point of view, mass is energy or energy is mass, which was discovered by Einstein almost a century ago [1]. And it is this basic fundamental law of physics that we have used for investigating the origin of time. Together with a huge energy explosion (i.e., Big Bang theory [10]), time is the igniter for the creation of our universe. As we know, without the existence of time, the creation of our universe would not have happened. As we have shown, time can be traded for space, but space cannot be traded for time. Our universe is in fact a temporal physical subspace, and it is continuously evolving or changing with time (i.e., $t > 0$). Although every temporal subspace is created by time (and substances), it is not possible for us to trade any temporal subspace for time. Since every physical substance has a life, our universe (a gigantic substance) cannot be excluded. With reference to the report from a recent Hubble Space Telescope observation [11,12], we are capable of viewing galaxies about 15 billion light years away and have also learned that our universe is still by no means slowing down in expansion. In other words, based on our estimation, our universe has still not reached its half-life. As we have shown, time ignited the creation of our universe, yet the created physical substances presented to us the existence of time.

One more thing I would note that why the speed of time moves forward at a constant velocity such as 1 second, 2 second, … . It is because that our universe was created with speed of light: from the relativity theory to time varying energy equation of Einstein shown in Eq. (1.15). In which we see that, every subspace within our temporal universe has to be temporally synchronized with the time variation within our temporal universe. Otherwise the subspace cannot exist within our temporal universe. For which a question arises is that for time traveling within our universe. For example if one can surgically remove oneself from our universe and searching backward for the same date of today of last year, one will find that the date of last year of today has already moved away. On the other hand, if one is forward searching for the same date of today in next year, one will not be able to find it because the date of this year today for the next year has not arrived yet, since time is a dependent variable coexisted with subspace.

1.13   Essence of Our Temporal (i.e., $t > 0$) Universe

In view of preceding discussion, we see that, our universe is a time-invariant system (i.e., from system theory stand point), as in contrast with an empty space, it is a not a time-invariant system and it is a timeless or on-time space. For which we see that; any timeless solution cannot be directly implemented within our universe. Since science is a law of approximation and mathematics is an axiom of absolute certainty, using exact math to evaluate inexact science cannot guarantee its solution exists within our temporal (i.e., t > 0) universe. One important aspect of temporal universe is that one cannot get something from nothing. There is always a price to pay; every piece of temporal subspace (or every bit of information [15]) takes an amount of energy (i.e., ΔE) and a section of time (i.e., Δt) to create. And the subspace [i.e., f(x, y, z; t), t > 0] is a forward time-variable function. In other words, time and subspace are coexisted or mutually inclusive. This is the boundary condition and constraint of our temporal universe [i.e., f(x, y, z; t), t > 0], in which every existence within our universe has to comply this condition. Otherwise it is not existed within our universe, unless new law emerges since all laws are made to be broken. Thus we see that any emerging science has to be proven existed within our temporal universe [i.e., f(x, y, z; t), t > 0]. Otherwise it is a fictitious science, unless it can be validated by repeated experiments.

In mathematics, we see that the burden of a postulation is first to prove that there exists a solution and then search for a solution. Although we hardly have had there is an existent burden in science, we need to prove yet that a scientific postulation exists within our temporal universe [i.e., f(x, y, z; t), t > 0]; otherwise it is not real or virtual as mathematics does; for example, the superposition principle in quantum mechanics in which we have proven [16] that it does not exist within our temporal universe (i.e., t > 0), since Schrödinger's quantum mechanics is timeless as mathematics does.

There is however an additional constraint as imposed by our temporal universe is that the affordability. As we have shown, everything (e.g., any physical subspace) existed within our universe has a price tag, in terms of an amount of energy ΔE and a section of time Δt (i.e., ΔE, Δt). To be precise, the price tag also includes an amount of "intelligent" information ΔI or an equivalent amount of entropy ΔS (i.e., ΔE, Δt, ΔI) [17]. For example, creation of a piece of simple facial tissue will take a huge amount of energy ΔE, a section of time Δt and an amount information ΔI (i.e., equivalent amount of entropy ΔS) to create. We note that, on this planet Earth, only human can make it happen. Thus we see that, every physical subspace (or equivalently substance) within our universe has a price tag (i.e., ΔE, Δt, ΔS), and the question is that can we afford it?

It may be interesting to know that why the speed of time is a constant-dependent variable. Since time is coexisted with substance (i.e., our temporal subspace), the speed of time is dictated by the speed of light with the creation of our temporal universe as can be seen from the differential form Einstein's energy equation in Eq. (1.10).

Furthermore, two of the most important pillars in modern physics must be the Einstein's Relativity Theory [1] and the Schrödinger's Quantum Mechanics [13]. One is dealing with very large objects (e.g., universe) and the other is dealing with very small particles (e.g., atoms). Yet, there exists a profound connection between them, by means of the Heisenberg's Uncertainty Principle [14]. In view of the Uncertainty Relation, every temporal subspace takes a section of time Δt and an amount of energy ΔE to create. Since we cannot create something from nothing, everything needs an amount of energy ΔE and a section of time Δt to make it happen. And my inquiry is that what would be the needed amount of energy ΔE and time Δt for the creation of a unit quantum limited subspace (QLS)? By referring to the Heisenberg Uncertainty limit, it is given as,

$$\Delta E \cdot \Delta t = h \qquad\qquad (1.25)$$

Figure 1.4 A quantum limited subspace (QLS), where r = c Δt is the radius of the subspace, c is the speed of light, $\Delta t = 1/\Delta v$ and v is the quantum state bandwidth.

where h is the Planck's constant. We see that every QLS is limited by ΔE and Δt, in which the h region limits the trading between ΔE and Δt. Since every temporal subspace can be determined by the speed of light, we see that the size of a QLS can be determined by its radius $r = c \Delta v$, where Δv is quantum state bandwidth, as depicted in Figure 1.4, in which we see that the size is proportional to Δt.

Since every temporal subspace is created by ΔE and Δt, and $\Delta E = \Delta m \cdot c^2$, we see that the radius of the QLS can be shown as,

$$r = c \cdot h / \Delta E = c / \Delta v \qquad (1.26)$$

where $\Delta E = h\ \Delta v$ is the particle's quantum state energy. Since the velocity of light is $c = 3 \times 10^8$ meter per second, the size of the QLS can be very large. Notice that, within the QLS complex amplitude information-transmission can be exploited [18], where a communication space that has not fully exploited yet.

1.14   *Boundary Condition of Our Universe*

It is very consequential to set up a fundamental boundary condition for our temporal universe for any scientific solution to comply; otherwise the solution may lead us to fictitious science that is not existed within our temporal universe. In view of the creation of our universe, we see that the fundamental boundary condition of any emerging science has to comply: the dimensionality and the temporal causality (i.e., $t > 0$) condition of our temporal universe. In other words every subspace (i.e., science) has to be dimensional, temporal, and complies with the causality constraint (i.e., $t > 0$). Otherwise the subspace (i.e., solution) is a virtual subspace not existed within our temporal universe. For instance, the speed of time within our universe is dictated by the speed of light, of which our universe was created with the speed of light. Even by a tiny fraction of a second slower or faster than the time-speed of our universe, the subspace (i.e., solution) cannot exist within our universe. In which we see that any analytical solution has to be dimensional, temporal, and causal that will guarantee the solution is existed within the boundary condition of our temporal universe. As you have seen that without the imposition of this fundamental boundary condition, fictitious sciences have already emerged all over in our scientific sectors, for example, the paradox of Schrödinger's cat [18], since 1935. Similarly the "instantaneous and simultaneous" multi-quantum states phenomena as promised by the Schrödinger's fundamental principle of superposition are actually not existed within our universe. As you have seen, the paradox of Schrödinger's cat has been debated by Einstein, Bohr,

Schrödinger, and many world renowned physicists for over 85 years since its disclosure at a Copenhagen forum in 1935, and still debating. If we have had the fundamental boundary condition at that time, then we would not have had produced a great number of fictitious sciences that was promised by the principle of superposition [18,19].

One of the important aspects of mathematics is the symbolic representation, by which complicated scientific consequence may be symbolically represented. This is precisely the reason; all the laws of science are point-singularity approximated. This is a scientific representation by no means can be used within our temporal (i.e., $t > 0$) universe. For example, take the famous equation of Einstein's energy equation as given by,

$$\varepsilon \approx mc^2 \tag{1.27}$$

which is a point-singularity approximated formula and has no dimension, no coordinate, and no time. In fact it is a timeless (i.e., $t = 0$) equation. If we treated this equation as analytical solution, we see this equation firstly is not a temporal equation, which cannot be directly used within our temporal universe. To comply with the temporal-causality condition, we can first transform the equation into a time-domain partial differential form and then put a constraint of $t > 0$ on it as given by:

$$\frac{\partial \varepsilon(t)}{\partial t} = c^2 \frac{\partial m(t)}{\partial t}, \ t > 0 \tag{1.28}$$

Then we see that this equation has transform into a time-dependent or temporal equation that satisfies the causality constraint of our temporal (i.e., $t > 0$) universe, and it can be used within our universe. This example would help us understanding the direct implementation of an analytic solution within our universe; firstly it has to comply with the temporal-causality constraint. In which we see that analytical science is mathematics, but mathematics is not "necessarily" equals to science. And the necessarily part implies the temporal-causality condition.

1.15   Time is a Discrete or Smooth Variable?

It is always an intriguing question to find out that time is a discrete or a continuous variable? To answer this intriguing inquiry; I will show that time is in fact a continuous and spatial dependent variable, as our temporal ($t > 0$) universe were created by Einstein's energy equation. In which we have shown time is subspace and subspace is time.

Time has been accepted as a discrete granular variable is largely dependent upon the finite particle standpoint by paricle physicists. For example: within particle physics regime they believed that all the matters or substances within our universe were created by elementary particles (i.e., no matter how small the particle is). For which they see that our universe must be granular or discrete. Since time is an dependent variable on its subspace, it is very convincing to discern that time is discrete and granularity.

However as we know that physical particles cannot exist within an empty space; in which we see that no matter how small particles are they have to be embedded within a temporal (t > 0) space, that is not empty. Therefore time "cannot" be a discrete variable. With reference to this physical evidence, once again the myth of science within our temporal (t > 0) universe is very incomprehensible for which we should be not settled that particle physics is the final destiny of micro space within our universe. Yet, it will take another level of scientific abstraction to figure it out, where those undetectable substances are; for example such as dark matters, dark energy, permittivistic (i.e., ϵ), permiable (i.e., μ) substances and many others, which includes the substance that allows gravitation field to be stored within our universe are.

Nevertheless, as we accepted physical particles cannot be embedded in a timeless (t =0) empty space, it tells us that it should always have substances between or among the particles; everlasting existed within our universe. And those substances have no rigged form as particles do; yet have fill up all the spaces and gaps within our universe and beyond our universal boundary. One of such substances is currently undetectable permittivistic (i.e., ϵ) and permeable (i.e., μ) media that have been known to be existed within our universe spaces. Otherwise electromagnetic waves cannot propagate within the huge universal space and this substance is known to be existed beyond the horizon of our universe; otherwise electro-magnetive waves would not able to propagate within our cosmological space and beyond the boundary of our universe. Once again we see that; our universe cannot be a granular subspace as particle physics suggested. In which we see that time has to be a smooth and continuous dependent variable, even within microparticle space enviroment. In view of our assessment of time it tells us that; it actually exists a permeable media or particle-less substance within our universe, which is waiting for us to discover!

Since speed of light is dependent on the permissive-permeable (i.e., ϵ, μ) media, it is in principle possible to artificially develop substances that their permittivity and permeability smaller that vacuum (i.e., < 1, μ < 1), in which the speed of light can push beyond the current speed, which remains to be developed.

1.16   We Are Not Alone?

Within our universe, we can easily estimate there were billions and billions of civilizations that had been emerged and faded away in the past 15 billion light years. Our civilization is one of the billions and billions of current consequences within our universe, and it will eventually disappear. We are here, and will be here, for just a very short moment. Hopefully, we will be able to discover substances that travel well beyond the limit of light before the end of our existence, so that a better observational instrument can be built. If we point the new instrument at the right place, we may see the edge of our universe beyond the limit of light. We are not alone with almost absolute certainty. By using the new observational equipment, we may find a planet that once upon a time had harbored a civilization for a period of twinkle thousands of (Earth) years.

1.17   Remarks

We have shown that time is one of the most intriguing variables in the universe. Without time, there would be no physical substances, no space and no life. With reference to Einstein's energy equation, we have shown that energy and mass can be traded. In other words, mass is equivalent to energy and energy is equivalent to mass, for which all mass can be treated as an energy reservoir. We have also shown that a physical space cannot be embedded in an absolute empty space or a timeless (i.e., t = 0) space and it cannot even have any absolute empty subspace in it. In reality, every physical space has to be fully packed with physical substances (i.e., energy and mass). Since no physical space can be embedded in an absolute empty space, it is reasonable to assume that our universe is a subspace within a more complex space, which remains to be found. In other words, our universe could have been one of the many universes outside our universal boundary, which comes and goes like bubbles. We have also shown that it takes time to create a physical space, and it cannot bring back the time that has been used for the creation. Since all physical substances exist with time, all physical spaces are created by time and substances (i.e., energy and mass). This means that our cosmos was created by time and a gigantic energy explosion, in which we see that every substance coexists with time. That is, without time, the creation of physical substances would not have happened. We have further noted that our universe is in a temporal space, and it is still expanding based on current observation. This shows that our universe has not reached its half-life yet, as we have accepted the Big Bang creation. And noted, we are not alone with almost absolute certainty. Someday, we may find a planet that once upon a time had harbored a civilization for a period of light years. We have further shown that the burden of a scientific postulation is to prove that there exists a solution within our temporal universe [i.e., f(x, y, z; t), t > 0]; otherwise it is not real or virtual as mathematics does. And we have also shown that the speed of time is dictated by the speed of light which is the igniter for the creation of our universe. We have shown a fundamental boundary condition for our universe, such that any analytical solution needs to comply with this boundary condition.

Finally, I would like to take this opportunity to say a few words on behalf of Professor Stephen Hawking, who passed away last year on March 14, 2018. Professor Hawking was a world renowned astrophysicist, a respected cosmic scientist, and a genius. Although the creation of temporal universe was started with the same root of Big Bang explosion, it is not a sub-space of Professor Hawking's universe. You may see from the preceding presentation that the creation of temporal universe is somewhat different from Hawking's creation. One of the major differences may be at the origin of Big Bang creation. My temporal universe was started with a Big Bang creation within a "non-empty" space, instead of within an empty space which was normally assumed. For which our universe is a sub-universe of a larger universe (i.e., multi-universes) which remains to be found. I have also shown that time and subspace is coexisted. Every subspace (i.e., substance) is created by energy and time, but the subspace cannot bring back the time that has been used for the creation since time is a forward-dependent variable. In other words every temporal subspace and time are interdependent; in which time cannot bring back the subspace that had been created. Similarly, the created subspace cannot bring back the section of time that has been used for the creation. In short we have seen that time is a smooth and "continuous" variable, since our universe is fully compacted with substance. And time is real and it is not an illusion.

References

1. A. Einstein, *Relativity, the Special and General Theory*, Crown Publishers, New York, 1961.
2. F. T. S. Yu, "Gravitation and Radiation," *Asian J. Phys.*, vol. 25, no. 6, 789–795 (2016).
3. A. Einstein, "Zur Elektrodynamik bewegter Koerper," *Annalen der Physik*, vol. 17, 891–921 (1905).
4. J. D. Kraus, *Electro-Magnetics*, McGraw-Hill Book Company, New York, 1953, p. 370.
5. M. Bartrusiok, *Black Hole*, Yale University Press, New Haven, CT, 2015.
6. G. O. Abell, D. Morrison, and S. C. Wolff, *Exploration of the Universe*, 5th ed., Saunders College Publishing, New York, 1987, pp. 47–88.

7. F. T. S. Yu, "Time: The Enigma of Space," *Asian J. Phys.*, vol. 26, no. 3, 143–158 (2017).
8. L. Amendola and S. Tsujikawa, *Dark Energy: Theory and Observation*, Cambridge University Press, Cambridge, 2010.
9. G. Bertone, ed., *Particle Dark Matter: Observation, Model and Search*, Cambridge University Press, Cambridge, 2010.
10. M. Bartrusiok and V. A. Rubakov, *Introduction to the Theory of the Early Universe: Hot Big Bang Theory*, World Scientific Publishing, Princeton, NJ, 2011.
11. J. O. Bennett, M. O. Donahue, M. Voit, and N. Schneider, *The Cosmic Perspective Fundamentals*, Addison Wesley Publishing, Cambridge, MA, 2015.
12. R. Zimmerman, *The Universe in a Mirror: The Saga of the Hubble Space Telescope*, Princeton Press, Princeton, NJ, 2016.
13. E. Schrödinger, "An Undulatory Theory of the Mechanics of Atoms and Molecules," *Phys. Rev.*, vol. 28, no. 6, 1049 (1926).
14. W. Heisenberg, "Über den anschaulichen Inhalt der quantentheoretischen Kinematik und Mechanik," *Zeitschrift Für Physik*, vol. 43, 172 (1927).
15. D. Gabor, "Communication Theory and Physics," *Phil. Mag.*, vol. 41, no. 7, 1161 (1950).
16. F. T. S. Yu, "The Fate of Schrodinger's Cat," *Asian J. Phys.*, vol. 28, no. 1, 63–70 (2019).
17. F. T. S. Yu, "Science and the Myth of Information," *Asian J. Phys.*, vol. 24, no. 24, 1823–1836 (2015).
18. K. Życzkowski, P. Horodecki, M. Horodecki, and R. Horodecki, "Dynamics of Quantum Entanglement," *Phys. Rev. A*, vol. 65, 012101 (2001).
19. T. D. Ladd, F. Jelezko, R. Laflamme, C. Nakamura, C. Monroe, and L. L. O'Brien, "Quantum Computers," *Nature*, vol. 464, 45–53 (March, 2010).

chapter two

Temporal Universe and Second Law of Thermodynamics

2.1 Second Law of Thermodynamics, a Revisit

One of the most intriguing laws in science must be the law of entropy. So what is entropy? The word entropy was originally coined by Clausius in 1855 [1]. He may have intended to be used as a negative effect that opposites of entropy and it has been found very useful in discussing from the negative-entropy (neg-entropy) perspective in science [2]. Historically, the importance of the neg-entropy concept was first started by Tait [3], a close associate of Kelvin [4]. And the increase in entropy has been treated as the degradation of energy by Kelvin [4]. Nevertheless neg-entropy represents the quality or grade of energy in an isolated system and must always decrease or remain constant. As in contrast with the positive-entropy in an isolated system, the entropy will always increase or remains constant [5]. In other words, the energy within an isolated system will always be degraded or maintains the same, in which we see that the concept of entropy is directly related with the degradation of energy. Nonetheless, what influences the entropy to increase and to remain constant is still remaining a mystery for the learners of the second law, particularly for the one who has no basic knowledge of thermodynamics.

Let us start with a composited temporal universe diagram [6,7] as shown in Figure 2.1, in which we see that every subspace is a temporal subspace within the universe. In view of the law of energy conservation, we see that energy within our universe is continuously degrading as her boundary expands at the speed of light. Thus we see that the entropy increase within every subspace in our universe is actually influenced by the entropy increase of the entire universe expansion with time.

Since every temporal subspace needs an amount of energy and time to create [6,7], this means that every subspace is not empty and it is temporal. Of which we see that any subsystem within our universe takes an amount of energy and a quantity of time (i.e., ΔE and Δt), or energy, time, and entropy (i.e., ΔE, Δt, and ΔS) to create. Hence, every isolated subspace can be described by a space and time representation, as given by

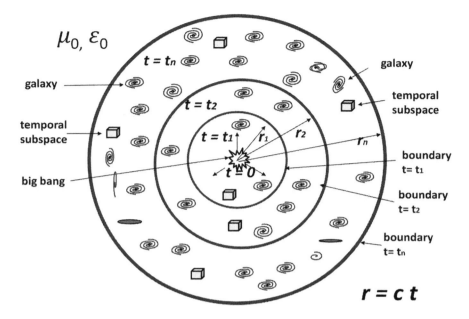

Figure 2.1   A composite diagram of our temporal universe.

$$f(x, y, z; t), t > 0 \tag{2.1}$$

where t is a forward time variable, $x = c{\cdot}t$; $y = c{\cdot}t$ and $z = c{\cdot}t$ are the spatial dimensions; and c is the speed of light.

We note that any timeless subspace cannot exist in our temporal universe or vice versa. And also every temporal subspace has to be compacted with substances (i.e., energy, mass, or both), as can be seen from the creation of our temporal universe in the following divergent energy vector description [6,7] as given by,

$$\frac{\partial \varepsilon}{\partial t} = -c^2 \frac{\partial m}{\partial t} = \nabla \cdot s \tag{2.2}$$

where ε is the energy, m is the mass, c represents the velocity of light, t is the time variable, $(\nabla\cdot)$ represents a divergent operator, and S represents the radiant energy vector. In view of the description of Eq. (2.2), one may see that how our universe was created from a big bang of huge energy explosion [8], in which the equation shows the transformation from a dimensionless (i.e., point singularity) presentation of energy explosion to a space-time description. This precisely shows that how our universe, as well as all of the temporal subspaces (i.e., space-time), were created. In Figure 2.1, we also see that our temporal universe is a bounded subspace and the boundary is still expanding at the speed of light based on our current observation [9]. Since our universe is a bounded temporal space, it has to be embedded in a more complex space, which remains to be found.

2.2   Second Law within Temporal Universe

Let us now look at the traditional second law as stated by [3]: "Entropy in an isolated system, the entropy will always increase with time or remains constant." This statement implies that the hypothetical isolated system is physically isolated from our universe and it is not a subsystem (or subspace) within our universe. Besides the physical isolation, there are a couple of fundamental questions that remain to be clarified.

Firstly, with what justification for the second law to assume that the entropy in an isolated system will always increase? Or equivalently, why energy within the isolated system will always degrade? Secondly, if the isolated system of the second law exists within our universe, then what influence the entropy in the isolated system to increase with time? Furthermore, why the entropy in the isolated subspace has to be "remains constant"? These are the myths of the second law; although the second law has been well accepted, yet puzzled scientists and engineers for over one and a half centuries.

The first question is that the isolated subspace has been a low-entropy subspace, as relatively with respect to the surrounding environmental entropy. In Figure 2.2, we show that an isolated lower-entropy subspace Q_0 is the subspace within a higher-entropy subspace Q_1. Since high energy flows to low energy, it is trivial to see that entropy in the lower-entropy subspace Q_0 has to increase with time. In other words, energy degrading in Q_0 subspace is faster than its surrounding subspace Q_1, which is primarily due to the energy degradation of the entire temporal universe with time.

Let me further note that, as the universe expands, the subspaces near the edge of the boundary expand faster than the subspaces closer to the center of our universe, as illustrated in Figure 2.3. This is precisely the reason why the entropy in any isolated subspace will continue to increase, although at different rates. We also see that the increasing entropy in every subspace will not come to standstill or stop, since the entropy increase in every subspace is affected by the expansion of the entire universe with time. For example, the speed of the subspaces closer to the edge of the boundary would be closer to the speed of light; we would anticipate the increase of entropy in those subspaces closer to the boundary would be faster than those subspaces near the center of the universe. The laws of science within those subspaces may

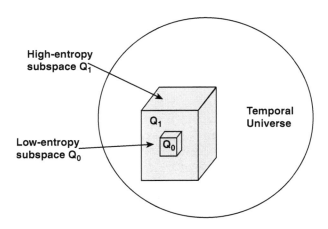

Figure 2.2   A lower-entropy subspace Q_0 that is within a higher-entropy subspace Q_1.

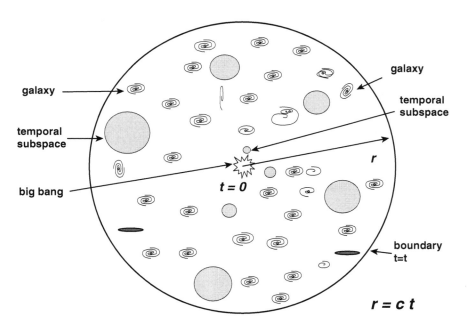

Figure 2.3   A temporal universe with her subspaces.

be different from ours and as well their relativistic times will not be the same as ours for those subspaces closer to the edge of our universe!

As we have noted, all laws were created to modify or to break, by which the second law has to be modified so that the myths of the law can be more precisely stated, as influenced by the entropy of entire universe changes. Thus, the modified version of the second law is written as: "Entropy in an isolated low-entropy subspace will always increase with time or maintains at a relatively lower-rate increasing as influenced by the entropy increases of the entire temporal universe with time."

This modified second law eliminates the myths of the law, and it is more precisely stated that, treated as the isolated system, a subspace within our temporal universe shows that entropy increase within the isolated system is influenced by the increase of entropy due to the entire universe. Nevertheless, the essence of the second law is that work can be performed between two subspaces if and only if there is an entropy imbalance between them, in which it shown that the second law prevails as a perpetuate motion machine cannot be realized within any balanced entropy isolated temporal subspace.

2.3   Second Law and Information

We assert that an isolated subspace within our universe can provide infor-mation if and only if it is a low-entropy subspace, as compared with the neighboring entropy. Let us consider N possible outcomes of an information subspace. If these N possible outcomes are assumed equi-probable, then the average amount of information provided by the source is given by

$$I_0 = \log_2 N \text{ bits/outcome} \tag{2.3}$$

Then a question is asked: What would be the amount of information needed for the deduction of N possible outcomes to lower M equi-probable possible outcomes within this subspace? Since $N > M$, the amount of information needed is apparently given by,

$$I = \log_2 N - \log_2 M = \log_2(N/M) \text{ bit/outcome} \tag{2.4}$$

Let us now look at a non-isolated temporal subspace in which the complexity of its N equip-probable states is established a priori. If the N equip-probable states, as time goes by, is reduced to M equip-probable states, then the entropy for the N and M equip-probable states can be written by Boltzmann–Planck equation [10] as respectively given by;

$$S_0 = k \, ln \, N \tag{2.5}$$

$$S_1 = k \, ln \, M \tag{2.6}$$

where k is Boltzmann's constant and ln represents the natural logarithm. Since $N > M$, the initial entropy S_0 is greater than the later entropy S_1 as given by,

$$S_0 > S_1 \tag{2.7}$$

By the virtue of the second law, we see that the non-isolated subspace cannot be isolated, because entropy will always increase in an isolated low-entropy subspace. In order for the entropy to decrease within a non-isolated subspace, an amount of information has to be provided to the subspace. However, the amount of information of I must be furnished by some external source, which is outside the non-isolated subspace. Thus we see that the decrease in entropy in the non-isolated subspace is apparently due to the amount of information provided (i.e., external source) into the non-isolated subspace. And this amount of entropy decrease is equaled to the amount of external information I provided, as given by,

$$\Delta S = S_0 - S_1 = -kI \, ln \, 2 \tag{2.8}$$

Or equivalently,

$$S_0 = S_1 - kI \, ln \, 2 \tag{2.9}$$

where $S_0 > S_1$. In which we see that the condition for a non-isolated subspace to decrease its entropy if and only if an amount of information is introduced into the subspace is proportional to the decrease in entropy ΔS within the non-isolated subspace. This equation in fact shows one of the important connections between the second law and information. The significance of Eq. (2.8), or equivalently Eq. (2.9), can be seen as trading entropy with information. Otherwise, information would be very difficult to be applied in science, because entropy is a well-accepted quantity in science.

However, the preceding non-isolated subspace does not guarantee a physical subspace yet. In order to be a physical subspace, it has to actually exist within our temporal universe, for which the non-isolated subspace has to be a temporal (i.e., $t > 0$) subspace within our universe, since every subspace within our temporal universe is affected by increasing the entropy of the entire universe, in which the non-isolated subspace cannot be the exception.

Then, what would be the amount of information required for reducing the entropy within a non-isolated subspace in our temporal universe? The answer is that, again it can be derived from an external source to make it happen, as we have shown earlier. To show how it works, this time we isolate the non-isolated subspace with an external information subspace of I, by which the new isolated subspace (i.e., the entire subspace) is a subspace within our temporal universe. Thus, as time goes by, the entropy of the entire subspace will increase or maintains at the same rate, as affected by the increasing entropy of our universe, as given by,

$$\Delta S = S_0 - S_1 = -kI \, ln \, (2) \; + \delta \cdot \Delta t \qquad (2.10)$$

where δ is the rate of entropy increase affected by the entire universal increases with time and Δt is the time interval of consideration. We see that Eq. (2.10) is essentially the same as Eq. (2.8), except with a remainder term $\delta \cdot \Delta t$. And this term needs a special mention, as compared with the overall entropy increase of the entire isolated subspace, which can be treated a small value for the time consideration. We can ignore this remainder term for simplicity in our discussion. Thus we see that the net entropy increased over time, without the inclusion of the remainder term $\delta \cdot \Delta t$, can be approximated by,

$$S' \; = \; S_1 - \delta \cdot t \approx S_1 \qquad (2.11)$$

Therefore the net entropy increased excluding the remainder term can be written as,

$$\Delta S' \; = \; S_0 - S' \approx - \, kI \, ln \, (2) \qquad (2.12)$$

Or equivalently we have,

$$S' \approx S_0 + kI \, ln \, (2) \qquad (2.13)$$

which is approximately equal to the result that we have obtained in Eq. (2.9), and it is similar with the same conclusion as based on original assumption, that is, the physically isolated system is not a subspace within the universe. In other words, the traditional second law is correct, but it was the way it stated which causes some inconsistencies. Hopefully this modified second law would minimize the ambiguous inscription and the myth if we treated the isolated system as a subspace within our temporal universe. For example, as stated by the traditional second law, entropy within the isolated system to increase or "remains constant" is kind of mystery, since the isolated subsystem has to be a temporal subspace within the universe, and added the entropy increases within every temporal subsystem will never stop. It is in fact continuing to increase but in a relatively slower pace as affected by the entropy increases of the entire universe with time. It has also been asserted that entropy increases within the isolated system cannot remain constant; otherwise the isolated system would be a non-temporal subspace. As we see, any non-temporal isolated system must be a "timeless" system, but a timeless subspace cannot exist within a temporal space. In other words, if entropy of an isolated system is capable to remain constant, then the isolated system cannot be a temporal system, for which the assumed isolated system has to be a subspace within our temporal universe.

2.4 Trading Entropy with Information

In this section, we will show a profound relationship between information and entropy, in which we will show that every bit of information acquired is not free; it is paid by an external source in which the entropy of the source is reducing. As referenced to the second law, when we isolate the entire system (i.e., includes an information source), for any further evolution within the entire system, the entropy will increase with time as given by,

$$\Delta S' \approx (S_o - kI \, ln \, 2) > 0 \qquad (2.14)$$

In which we see that, any further increase in entropy $\Delta S'$ can be due to ΔS_0 or ΔI, or both. Although in principle it is possible to distinguish the changes in ΔS_0 and ΔI separately, in some cases, the separation of the changes due to ΔS_0 and ΔI may be difficult to discern.

It is interesting to note that if the initial entropy S_0 of the isolated subspace corresponds to some complexity of a structure but not the maximum, and if S_0 remains unchanged ($\Delta S_0 = 0$), then after a certain free evolution without the influence of external sources, from Eq. (2.14) we will have

$$\Delta I < 0. \qquad (2.15)$$

Since $\Delta S_0 = 0$, with reference to Eq. (2.14), the changes in information ΔI is very smaller, or decreasing. The interpretation is that when we have no knowledge of the entire subspace's complexity, the entropy S_0 is assumed to be maxima (i.e., the equi-probable case). Hence, the information provided by the subspace structure is maxima. Therefore, $\Delta I < 0$ is due to the fact that in order to increase the entropy of the subspace, $\Delta S_0 > 0$, a certain decrease in information is needed. In other words, information can be provided or transmitted (a source of neg-entropy) only by increasing the entropy of the subspace. However, if the initial entropy S_0 is at a maximum condition with respect to the surrounding entropy (i.e., $\Delta I = 0$), the subspace cannot be used as a source of neg-entropy. Nonetheless, we have shown that quantity of entropy and the amount of information can be traded, as symbolically described by,

$$\Delta S' \rightleftharpoons \Delta I \qquad (2.16)$$

In which we see that information can be obtained only by increasing the entropy from a physical device, which is a lower-entropy device as compared with the surrounding entropy. In other words, the increase in entropy from a physical system can be used as a source of neg-entropy to provide information, or vice versa. In fact, work done or energy can be shown to be related to information, as given by;

$$\Delta W = \Delta Q = T \ \Delta S = IkT \ ln \ 2 \qquad (2.17)$$

where W is work done, Q is the heat, and T is the thermal noise temperature in Kelvin. In which we see that, with higher thermal noise temperature, the amount of energy required for the transmission of information is higher.

In view of Eq. (2.16), I would stress that the amount of information (e.g., ΔI in bits) or equivalently amount of entropy (i.e., $\Delta S'$ in joules per Kevin) is the "cost" needed to obtain (e.g., to purchase) the bits of information but not equal to the information. For example, a book has N bits (or equivalent amount of entropy), but we have numerous numbers of books that have the same N bits of information. This is similar as: an apple is cost a dollar in which we see that a dollar can also purchase an orange or even a roll of toilet tissues.

It remains however a question to be answered: Is there a reversal second law as stated? "Entropy of an isolated subspace within our universe will decrease with time." The answer is yes, if the isolated subspace includes an energy convergent substance, for example such as a Black Hole [11].

2.5   Remarks

Since the discovery of entropy by Clausius in 1865, the myth of second law of thermodynamics has intrigued scientists and engineers for well over a century. This law could have been one of the most quoted scientific laws by physicists, chemists, engineers, and information scientists. However, all laws were created to be modified or to be broken, and the second law cannot be the exception. On the other hand, the important aspect of the creation of our temporal universe is that every subspace within our universe has to be temporal. Any evolution of our universe with time has a profound effect in her subspaces, which includes the second law. It is this reason of this chapter to make a modification of the second law so that it will be more precisely stated as consistent within our temporal universe. The new modified second law is stated as "Entropy in an isolated low-entropy subspace will always increase with time or maintains at a relatively lower-rate increasing as influenced by the entropy increases of the entire temporal universe with time." In addition, we have shown that, there existed a profound relationship between second law and information, in which we see that entropy and information can be traded. We have also stressed that, without the second law, information would be difficult to apply in science. The reason is that, entropy is a well-accepted quantity in science. We have further shown that, low-entropy subspace (or devise) can be used as a neg-entropy source to provide information, of which we shown that work done and energy are all related to information. In short, we note that the higher the thermal noise temperature, the higher the energy requires for information transmission.

Since the amount of information (or equivalent amount of entropy) has been frequently misinterpreted as the information, I have pointed that it is the "cost" in bits (or in joules per Kelvin) to obtain (i.e., to purchase) the information, but not the information.

References

1. R. Clausius, "Ueber verschiedene für die Anwendung bequeme Formen der Hauptgleichungen der mechanischen Wärmetheorie: vorgetragen in der naturforsch," *Gesellschaft den*, vol. 24, 46 (April 1865).
2. L. Brillouin, "The Negentropy Principle of Information," *J. Appl. Phys.*, vol. 24, 1152 (1953).
3. P. G. Tait, *Sketch of Thermodynamics*, Edmonston and Douglas, Edinburgh, 1868, p. 100.
4. L. Brillouin, *Science and Information Theory*, 2nd ed., Academic, New York, 1962.
5. F. T. S. Yu, *Optics and Information Theory*, Wiley-Interscience, New York, 1976, p. 80.
6. F. T. S. Yu, "Time: The Enigma of Space," *Asian J. Phys.*, vol. 26, no. 3, 143–158 (2017).
7. F. T. S. Yu, *Entropy and Information Optics: Connecting Information and Time*, 2nd ed., CRC Press, Boca Raton, FL, 2017, pp. 171–176.
8. G. O. Abell, D. Morrison, and S. C. Wolff, *Exploration of the Universe*, 5th ed., Saunders College Publishing, New York, 1987, pp. 47–88.

9. R. Zimmerman, *The Universe in a Mirror: The Saga of the Hubble Space Telescope*, Princeton Press, Princeton, NJ, 2016.
10. L. Boltzmann, "Über die Mechanische Bedeutung des Zweiten Hauptsatzes der Wärmetheorie," *Wiener Berichte*, vol. 53, 195–220 (1866).
11. M. Bartrusiok, *Black Hole*, Yale University Press, New Haven, CT, 2015.

chapter three

Science and the Myth of Information

3.1   Space and Information

Let us start with a bit of Information provided by a unit space. As we have seen, every physical space is temporal and it is created by energy and time (i.e., ΔE, Δt). Therefore, a physical space that we are dealing with is not empty; otherwise it would be an absolute empty, which does not exist within our universe. For which, we assume an information cell or a unit space as illustrated in Figure 3.1(a).

Virtual reality is an imaginary reality without the supports of the current law of science (e.g., mathematical abstract space), and the physical reality is supported by the laws of science. In order to make this unit cell a reality, one should first look into the time-space reality. Time is the most essential quantity in our physical science that governs all the laws of physics. Without time, there would be no space, no substance, and no life. Thus time must be the supreme law of science. In other words, all the laws of science cannot exist without time.

Let us show the unit cell as to becoming a real space. To do so, we would take the ultimate limit in the current laws of physics, the speed of light. With the imposition by speed of light c, each dimension can be described by $d = c \cdot \Delta t$, where Δt is time duration. Nonetheless, this unit space is still not yet a physical space, since it is absolutely empty or null. As we know that an absolute empty space cannot be embedded in a real space, we have to fill up the space with substance or substances. By filling up the space with a substance, then it is a real unit space within the law of science.

If we assume this unit cell provides a dual-level signal at a time (i.e., white or black color) shown in Figure 3.1(b), the information provided by this unit-cell is given by [1, 2],

$$I = \log_2 2 = 1 \text{ bit} \tag{3.1}$$

There is however a question remains to be asked: What would be the lowest dimension of a unit information cell? The lowest limit must be imposed by the Heisenberg uncertainty principle. In other words, it is

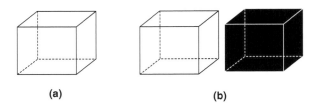

Figure 3.1   An information cell. (a) A virtual unit cell. (b) Binary information cell.

the time duration Δt and the speed of light c that set the size of the unit cell under the constraint of the current law of science.

Now, let us extend this unit information-cell to a cubic N×M×H information cells shown in Figure 3.2.

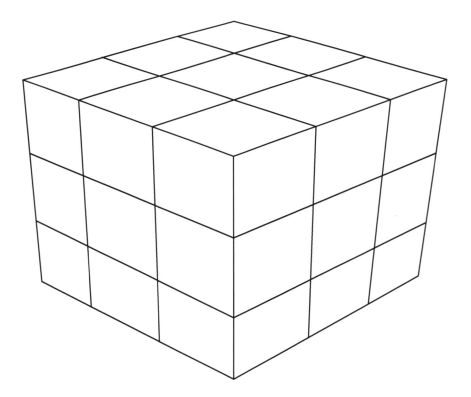

Figure 3.2   A cubic N×M×H information cells.

Then the information provided at a given time of this creature can be written as:

$$I = (N \times M \times H)bits, \qquad (3.2)$$

Again we have assumed that the probability occurrence of a black or a white cell is equally probable.

Now, change the problem slightly. If each unit cell is equally capable of giving one of W distinguishable gray-level signals at any given time, then the overall spatial information provided by this 3D cubic structure would be

$$I = (NxMxH) (\log_2 W) \text{ bits.} \qquad (3.3)$$

This is also in fact the amount of information (i.e., cost) needed to create this 3D cubic information structure, with the assumption of 100% certainty!

3.2   Time and Information

Vocal voice is a typical example of a temporal information or a time signal. Telegram (Morse code) signal is another example. A picture is an example of a space information or spatial signal. Needless to mention that, spatial and temporal signals can be interfaced or interweaved for temporal transmission! For instances, TV display is an example of exploiting the temporal information transmission for spatial information display, old-fashion movie sound track is an example of exploiting the spatial information for temporal information transmission, and a continuous running TV program is an example for spatial and temporal information transmission.

Let us now look at the information content of a time signal as depicted in Figure 3.3(a).

As we all know that, any temporal signal can be digitized in binary form for temporal transmission or time-signal transmission, as illustrated (i.e., 0 and 1) in Figures 3.3(b) and 3.3(c). This is precisely the binary or digital format we used in current communication and computer systems. However, most of the people including some engineers and scientists know how the digital system works, yet some of them may not know why we developed it.

Now, let us start with the major differences between the digital and analog systems as follows: Digital system operates in binary form (i.e., 0, 1), while analog system operates in analog form (i.e., multi-level). Digital system provides lower information content (e.g., one bit per

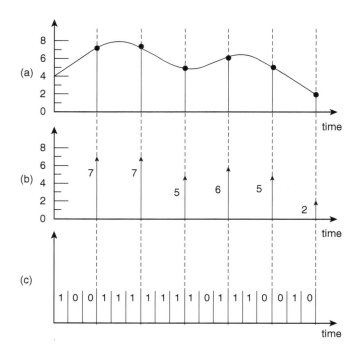

Figure 3.3   Analog to digital signal conversion. (a) Analog signal. (b) Sampled signal. (c) Digital signal.

level), while analog system provides higher information content (e.g., more bits per level) and others. Since the information content of a digital system is lower than an analog system, why does one go through all the troubles to transform the analog to digital and then to transform back to analog for the receiver? The answer is that by exploiting the transmission at velocity of light, which can carry a lot of temporal information at that speed. And it is precisely the price paid for the time transmission.

Since the major purpose of using digital transmission is for noise immunity, otherwise we will not use longer encoded digital transmission. The reason is that, in digital transmission, the signal can be easily repeated, as in contrast with analog signal transmission, it cannot. One may see that, after a few cycles of amplifications, an analog signal will be completely corrupted by noise; while in digital-signal transmission, the transmitted signal can be easily refreshed by means of a repeater. Thus, a digital signal can be transmitted over thousands of miles by a repeater or repeaters and the received signal is just as good as the original!

As an example, if one consecutively copies a compact disc or a DVD for many times, one would discover that the latest copy is just as good as the original one! Although the digitally transmitted signals strictly speaking are not real time, it appears to be very close to real time because of the light speed transmission! And this is precisely the price we paid for time transmission at the speed of light!

Since time is one of the most esoteric quantities in the myth of science, it has no weight and no volume, no beginning, and no end. It cannot stand still and go backward. And every physical substance is constantly in motion with time. The fact is that every piece of information can only convey with time. Thus, without the existence of time, all the information would not be able to convey and observe! In other words, every piece of information takes time and energy to be transmitted and it is not free. A trivial example is that when one is traveling at the speed of light (assuming that one could), one cannot even see himself with a mirror at the front of him, since there is no expenditure in time. Thus, we see that it takes time for a piece of information to convey.

3.3   Entropy and Information

In reality, our world and this universe are filled with information. Then there is a question that had been asking: What would be a connection between information and science? In other words, without a physical quantity to represent information, information would be difficult to apply and inability to connect with science!

One of the most intriguing laws in the Second Law of Thermodynamics must be the law of entropy, which was developed by Boltzmann as given in [3]:

$$S = k \ln N \tag{3.4}$$

where k is the Boltzmann constant, ln is the natural log, and N is the total number of the possible distinguishable states. In view of the information measure (e.g., in bits) is given by:

$$I = -\log_2 (p) \tag{3.5}$$

where p is the probability of a possible state. We see that there exists a profound connection between information and the law of entropy, in which both equations are statistical in nature and also were presented in logarithmic forms. It is now apparent that each amount of information is associated with an amount of entropy [4]. In other words, every piece of information is equivalent to an amount of entropy, a quantity that has been well accepted in science.

Let us now show that information can indeed be traded with entropy. Given a non-isolated physical system in which equal probability in complexity structure has been established, we further assume that the system has N possible outcomes and has been reduced to M states. Then the amount of entropy for N and M equi-probable states can be respectively written as:

$$S_0 = k \ln N \qquad (3.6)$$

and
$$S = k \ln M \qquad (3.7)$$

where $N > M$, and k is the Boltzmann's constant. For which we see that

$$S_0 > S_1 \qquad (3.8)$$

Since it is a non-isolated system and its entropy can be reduced, if and only if an amount of external-information is introduced into the system, we have:
$$\Delta S = S_1 - S_0 = -kI \ln 2 \qquad (3.9)$$

Or equivalently, preceding equation can be written as:

$$S_1 = S_0 - kI \ln 2 \qquad (3.10)$$

Thus, we see that, an amount of information I is needed for this reduction, which is proportional to the decrease in entropy ΔS. This is precisely the basic connection between entropy and information, of which we see that the information and entropy can be simply traded!

Moreover, if one isolates the entire system for consideration, as reference to the second law of thermodynamics, we see that for any further evolution within the system, its entropy should either increase or remain constant, as given by:

$$\Delta S_1 = \Delta(S_0 - kI \ln 2) \geq 0 \qquad (3.11)$$

where we see that any further increase in entropy ΔS_1 is due to ΔS_0 or ΔI or both. Although in principle it is possible to distinguish the changes in ΔS_0 and ΔI, in some cases the separation of the changes may be difficult to discern. We further note that, since the (isolated) system is not influenced by any external source i.e., $\Delta S_0 = 0$, we see that the changes in information would be negative or decreasing:

$$\Delta I \leq 0 \qquad (3.12)$$

Since the original entropy S_0 is assumed maximum (i.e., the equally probable case), $\Delta I \leq 0$ is due to increase in entropy of the system ($\Delta S_1 \geq 0$), a decrease in information is needed. In other words, information can be provided or transmitted (a source of neg-entropy) only by increasing the entropy of the system. However, if the initial entropy S_0 is at a maximum state, then $\Delta I = 0$ and the system cannot be used as a source of neg-entropy. For example, a dead battery (i.e., maximum entropy) cannot be used to provide information. This illustrates that entropy and information can actually be interchanged or simply traded, but at the expense of some external source of entropy, it is given as,

$$\Delta I \rightleftharpoons \Delta S \qquad (3.13)$$

Let me emphasize that this relationship is one of the most intriguing connection of information with science; otherwise information would not have any direct physical connection as applied to science.

3.4 Substance and Information

Every substance has a piece of information that includes all the elementary particles, basic building blocks of all elements, atoms, paper, our planet, solar system, galaxy, and even our universe! In other words, the universe is flooded with information (i.e., spatial and temporal) or the information fills up the whole universe. Strictly speaking, when one is investigating the origin of the universe, which includes time-space as well as the existence of life, the aspect of information cannot be absence. Then, one would ask: What would be the amount of information, aside the needed energy, required to create a specific substance? Or equivalently, what would be the cost of entropy to create it?

There are several kinds of spatial information or simply information needed to define before our discussion. From physical standpoint, there are two types of spatial information that I would define, namely, Static (or pro-passive type) and Dynamic (or pro-active type) Information. What I meant Static Information is primarily created by the consequences of Mother Nature's law of physics. For instance, an amount of information is needed to create an atom or an elementary particle. In other words, Static Information is not created by bio-evolutional process or man-made. Static information was mostly formulated by complicated nuclei and chemical reaction or interaction. And they are not created due to intelligent or human intervention!

Samples of Static Information such as information of an elementary particle, an electron, a proton, and various atomic particles (e.g., hydrogen atom, oxygen atom, uranium, gold, etc.) are typical examples. In fact, to actually produce those subatomic particles, aside from the extreme nuclei interaction (as we shall discuss later), an amount of information (or entropy) is required for a successfully creation. This process cannot be done by simple human intervention or by evolutional process, but by means of statistical consequences from nuclei and chemical interaction; similar to our universe, solar system and planets were formed.

On the other hand, Dynamic Information is mostly man-made or by biological (i.e., evolutional) processes. In fact the Dynamic Information that I intended to mention is not created by Mother Nature, but generated either by evolutional processes or by man-made. Nevertheless, Dynamic Information has two sub-categories: namely, Artificial and Self-organized Information. Artificial Information is the man-made information, while Self-organized Information is self-produced or by means of adaptation, such as information contents of DNA, gene, immune system, or a biological cell, and others.

Practically all of the things made by human are considered as Artificial, which include the tools and devices, such as watches, computers, battle ships, and automobiles that are typical examples. Nonetheless, each of the devices has its own signature associated with a specific amount of information. And this is the amount of information or artificial information needed, aside from time and energy, for the creation to happen. Again, this information needed for creation is not free and it will cost a lot of entropy.

3.5   Uncertainty and Information

One of the most fascinating principles in Quantum Mechanics [5] must be the Heisenberg's uncertainty principle as given in [6]:

$$\Delta x \cdot \Delta p \geq h \qquad (3.14)$$

where Δx and Δp are the position and momentum errors, respectively, and h is the Planck's constant. The principle implies that one cannot detect or observe the position and momentum (e.g., a particle) simultaneously, within an error of Planck's constant or h region. Needless to say that Heisenberg uncertainty relation can also be written in various forms [7] as follows:

$$\Delta E \cdot \Delta t \geq h \tag{3.15}$$

$$\Delta \nu \cdot \Delta t \geq 1 \tag{3.16}$$

where ΔE, Δt, and $\Delta \nu$ are energy, time, and spectral resolution, respectively.

We emphasize that the lower bound of $\Delta \nu \cdot \Delta t \geq 1$ offers a very significant relationship with information or equivalently equals to an information cell as shown by Gabor [8] and illustrated in Figure 3.4, where ν_m and T denote the frequency and time limits of a time-signal, in which we see that every bit of information can be efficiently conveyed or transmitted, if and only if it is operating under the constraint of the uncertainty principle (i.e., $\Delta \nu \cdot \Delta t \geq 1$). This relationship implies that the signal bandwidth should be either equal or smaller than the system bandwidth; i.e., $1/\Delta t \leq \Delta \nu$, in which we see that Δt and $\Delta \nu$ can be traded. It is however the unit region but not the shape of the information cell that determines the limit.

Similarly, with reference to the uncertainty relationship $\Delta E \cdot \Delta t \geq h$, we see that it is the unit Planck's region or h region sets the boundary and it is not the shape. Since Planck's region is related to the unit information cell as we have described, it is trivial to discern that every bit of information needs an amount of energy ΔE and an amount of time Δt to efficiently transfer, to create, and to observe, and again it is not free!

By referring to Figure 3.4, the total number of information cells can be written as,

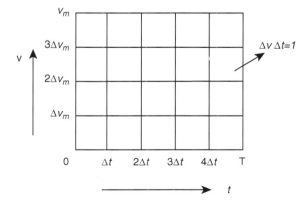

Figure 3.4 Gabor's information cell.

$$N' = (\nu_m/\Delta\nu)(T/\Delta t) \tag{3.17}$$

We note that the band-limited signal must be a very special type. For which the function has to be well behaved, it contains no discontinuity, no sharp angles, and has only rounded-off features. This type of signals must be analytic functions! And the shapes of the information cells are not particularly critical; it is however the unit area

$$\Delta\nu \cdot \Delta t = 1 \text{ (or } h \text{ region for } \Delta E \cdot \Delta t = h). \tag{3.18}$$

Furthermore, as suggested by Gabor, a set of Gaussian cosine and sine wavelets can be used in each cell as shown in Figure 3.5,

Needless to mention, if each information cell is accommodated with this set of elementary signals or wavelets for consideration, then the total number of information cells would be twice the amount, such as $N = 2N'$. Then the uncertainty principle can be respectively deduced to the following inequalities:

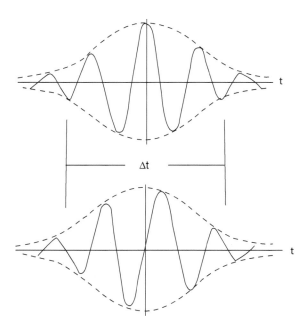

Figure 3.5 Gaussian envelop; cosine and sine wavelets.

$$\Delta x \cdot \Delta p \geq (1/2)h \tag{3.19}$$

$$\Delta E \cdot \Delta t \geq (1/2)h \tag{3.20}$$

$$\Delta \nu \cdot \Delta t \geq 1/2 \tag{3.21}$$

where the h or unit region has been reduced to a half.

To wrap up the uncertainty and information discussion, we provide a set of sound spectrograms, which shows uncertainty principle indeed holds, as depicted respectively in Figures 3.6(a) and 3.6(b).

Figure 3.6(a) shows that a wide-band sound spectrogram in which the time resolution Δt (i.e., time striation) can be easily visualized, but at the expense of finer spectral resolution $\Delta \nu$. On the other hand by viewing the narrow-band analysis in Figure 3.6(b), we see that finer spectral resolution $\Delta \nu$ can be achieved, with the expense of time striation Δt! Notice that these results are consistent with the uncertainty principle: one cannot observe fine spectral resolution $\Delta \nu$ and time resolution Δt simultaneously [9]!

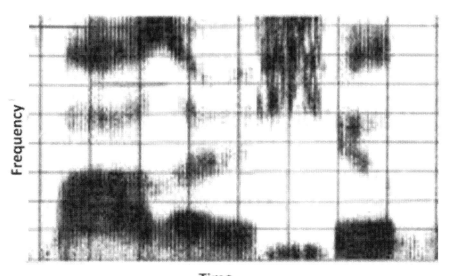

Time

Figure 3.6a Wide-band sound spectrogram.

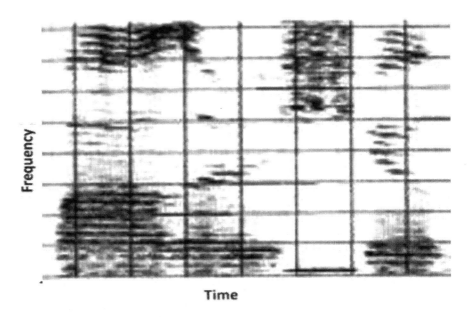

Figure 3.6b   Narrow-band sound spectrogram.

3.6   *Certainty and Information*

Uncertainty principle was derived under the assumption of invasive observation, in which the specimen under investigation was using an external illuminator. Although the principle is appeased not to be violated, it does not mean that within the certainty limit cannot be exploited. Thus, by reversing the inequality, we have,

$$\Delta t \, \Delta v \leq 1 \tag{3.22}$$

It is, however, reasonable to name this reversal relationship as "Certainty Principle," as in contrast with the uncertainty principle. This means that when the signal (e.g., the light beam) propagates within a time window Δt, the complex light field preserves a high degree of certainty! Thus, as the bandwidth Δv of the light beam (or signal) is decreasing, the signal property is self-preserving (i.e., unchanged) within a wider time window Δt, or vice versa! This is precisely the temporal coherence limit of an electro-magnetic signal (or light beam). If one multiplies the preceding certainty inequality with the velocity of light c, we have:

$$c\,\Delta t \leq c/\Delta v \tag{3.23}$$

This is the certainty distance (or coherence length) of a signal beam (or light source), as written by:

$$\Delta d \leq c/\Delta v \tag{3.24}$$

This means that within the coherence length Δd, the signal u(t) would be highly correlated with preceding time signal
$u(t + \Delta t)$, as expressed by

$$\Gamma_{12}(\Delta t) = \lim_{T\to\infty} \frac{1}{T} \int_0^T u_1(t)u_2^*(t + \Delta t)dt \tag{3.25}$$

where $\Delta t \leq 1/\Delta v$, * denotes the complex conjugate, μ_1 is the received signal, and u_2^* is the signal before Δt. This expression is named as Certainty Function, which is similar as the Mutual Coherence Function in coherence optics [10]. And the Degree of Certainty (i.e., Mutual Coherence) can be determined by the following equation:

$$\gamma_{12}(\Delta t) = \frac{\Gamma_{12}(\Delta t)}{[\{\Gamma_{11}(0)\ \{\Gamma_{22}(0)\ ]^{1/2}} \tag{3.26}$$

As regarded earlier, the shape of an information cell (i.e., $\Delta v \cdot \Delta t = 1$) is not a major issue, as long it is a unit cell. We stress that it is within this unit region that has not been fully exploited, such as applied to signal transmission, observation, processing, measurement, and imaging. Let us now show a couple of examples to illustrate that within the certainty region, $\Delta t\,\Delta v \leq 1$, complex wave front reconstruction can be indeed exploited as follows.

One of the successful applications within the certainty region (i.e., within the unit region) must be the application to wavefront reconstruction (i.e., holography) of Gabor [11]. We know that a successful holographic construction is dependent upon the coherence length (i.e., temporal coherence) of the light source. In other words, the light source controls the certainty relationship such that the object beam and the reference beam would be highly correlated (i.e., mutually coherence). Otherwise, the complex wavefront of the object beam would not be properly encoded on a photographic plate.

Another example is applying to synthetic aperture radar (SAR) imaging. The returned radar signal is required to be combining with a highly coherence local signal, so that the complex distribution of the returned radar wavefront can be synthesized on a square-law medium. Let us now provide the experimental results that were obtained within the certainty region as depicted in Figure 3.7.

Figure 7(a)   Hologram image.

Figure 3.7(a) shows an image that was reconstructed from a hologram, which was recorded within the coherent length or certainty distance (Δd) of a laser (i.e., light source). This guarantee that the object and reference beams be coherently encoded on a photographic plate. Figure 3.7(b) shows a radar imagery that was obtained from a synthetic aperture format [12], which was synthesized by a series of reflected radar wavefronts with a mutually coherent local signal. We further note that microwave radar has a very narrower bandwidth. In practice, its certainty distance (or coherence length) Δd could be over hundred thousands of feet!

3.7   Relativity and Information

Since every bit of observation is limited by the uncertainty relationship as given by:

$$\Delta \nu \cdot \Delta t \geq 1 \qquad\qquad (3.27)$$

Figure 7(b)   SAR image.

In which we note that its spectral resolution and time resolution can be traded. In other words, it is the unit cell or region, but not the shape that determines the boundary. For example, the shape can be an elongated rectangle shape, as long as it is limited by a unit region. In practice, time is one of the most mysterious elements in science; it can only move forward and cannot move backward. By means of Einstein's special theory, time may slow down somewhat, if one travels closer to the speed of light. Now, let us take the Time-Dilation expression from Einstein's special theory of relativity [13] as given by:

$$\Delta t' = \frac{\Delta t}{\sqrt{1 - v^2/c^2}} \tag{3.28}$$

where $\Delta t'$ is the dilated time window, Δt is time window, v is the velocity of the observer, and c is the speed of light. If we assume the observer is traveling at a velocity v and observing a standstill (i.e., $v = 0$)

specimen, then the observer would have a wider time window $\Delta t'$, instead of Δt. Then as applied to the uncertainty principle, we have,

$$\Delta \nu \cdot \Delta t' \geq 1 \qquad (3.29)$$

Since the dilated time window $\Delta t'$ (i.e., observation time window) is wider than Δt, i.e., $\Delta t' \geq \Delta t$, a finer spectral resolution limit $\Delta \nu$, in principle, can be obtained.

We further note that, as the velocity of the observer v approaches to the speed of light (i.e., $v \to c$), then $\Delta t' \to \infty$ the time window $\Delta t'$ would become infinitely large (i.e., $\Delta t' \to \infty$)! This means that the observer, in principle, can observe the specimen as long as he wishes, while the observer is traveling at the speed of light! In this case the observer would have, in principle, infinitesimal small (or finer) spectral resolution (i.e., $\Delta \nu \to 0$).

On the other hand, if the observer is standing still (i.e., $v = 0$) and is observing an experiment which is traveling at velocity v, then the time window for consideration is

$$\Delta t = \Delta t' \sqrt{1 - v^2/c^2} \qquad (3.30)$$

By substituting Δt in the uncertainty relation i.e., $\Delta \nu \cdot \Delta t \geq 1$, we see that a broader (i.e., poorer) spectral resolution $\Delta \nu$ would be obtained, since $\Delta t \leq \Delta t'$.

Again, if the velocity of the specimen is approaching the speed of light (i.e., $v \to c$), the observer would have no time to observe since $\Delta t \to 0$. And the spectral resolution would be infinitely large or extremely poor (i.e., $\Delta \nu \to \infty$)!

Two of the most important pillars in modern physics must be the Einstein's relativity theory and Schrödinger's quantum mechanics. And we have shown earlier, there is a profound connection between these two pillars, by means of Heisenberg's uncertainty principle, in which the observation time window can be altered by time dilation of Einstein!

Let us glimpse again the significances of the Heisenberg's uncertainty principle as written in following forms:

$$\Delta E \cdot \Delta t \geq h \qquad (3.31)$$

$$\Delta p \cdot \Delta x \geq h \qquad (3.32)$$

We note again, the constraints are not by the shape but by the quantity of the Planck's constant *or* h region. Since each h region is also related to an information cell, we see that the energy ΔE and the time resolution Δt can also be traded, as well as exchangeable between the momentum Δp and the position Δx variables. In other words, it is the Planck's region that limits the trading limit for ΔE and Δt, as well for Δp and Δx. However, in practice, it is the time window Δt more difficulty to manipulate, since time is constantly moving forward and cannot even slow down or stand still! Nonetheless, we have shown that the observation time window can be widened as the observer travels closer to the speed of light.

3.8   Creation and Information

We have noted that every substance has a distinctive signature or a piece of information. And this piece of information is equivalent to a cost of entropy. Since information and entropy can be traded, it is the cost of entropy needed to produce the equivalent amount of information. But, in reality, the cost of entropy is usually very excessive!

Now, let us go back to the 3D cubic structure. As illustrated in Figure 3.2, we had shown that the amount of information provided is given by:

$$I = (NxMxH) \, (\log_2 W) \text{ bites} \qquad (3.33)$$

And the equivalent amount or cost of entropy can be shown as:

$$S = (NxMxH) \, (\log_2 W) \, kln2 \qquad (3.34)$$

This is precisely the price paid or the minimum cost of entropy to create this cubic structure.

For simplicity of illustration, we let this cubic structure have a total mass of m, then what would be the minimum amount of energy required to create it? With reference to Einstein's total energy equation [13], i.e.,

$$E \approx mc^2, \qquad (3.35)$$

where m is the mass and c is the velocity of light. We see that this must be the minimum cost of energy needed if one wishes to reverse energy into mass, in which we assume that one can do it!

For example, what would be the cost of entropy to create a 3-unit binary-information cell shown in Figure 3.8?

If this 3-unit cell structure has a mass of one kilogram, what would be the energy required for one to make it happen?

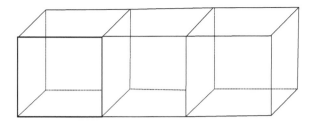

Figure 3.8   A 3-unit information cells.

First, it would require a minimum cost of energy about 1000 atom bombs dropped in Nagasaki at the end of World War II! Notice that this is the assumption that the one can reverse the total energy equation back to mass. Even though we assume one could reverse it, but it still not sufficient to create the 3-unit information cells. To do so he/she still needs the 3 bits of information (or minimum cost of entropy) to make it happen! And the cost of entropy needed can be calculated as shown by $S = 3\ kln(2)$.

We further note that the minimum cost of energy to create the 3-bit cubic-structure is a necessary condition but not sufficient yet to create the whole structure. And it is the 3-bit information or the cost of entropy $S = 3\ kln(2)$ needed to make it happen! And these 3 bits of information (or equivalent amount of entropy) is the sufficient condition to create the structure. Notice that the 3 bits of information could either come from Mother Nature or man-made. In this example, it is more likely coming from man-made!

3.9   Price Tags and Information

We stress again that every bit of information has a price tag or price tags! For instance, every bit of it is attached with a cost of entropy and it is limited by Δt and ΔE of the uncertainty principle. This means that every bit takes time and energy to transmit, to process, to record, to retrieve, to learn, and to create. And it is not free. Then a question is that: Can one afford to pay for it?

Let us assume a mechanical clock and its dissembled parts are respectively shown in Figures 3.9(a) and 3.9(b). Then, a question is asked: If we provide all the needed parts to a layman to reassemble it, what would be the amount of information needed or the cost of entropy required for him/her to reassemble it? And a follow-up question is that: How long will it take for him/her to actually assemble it?

First, since the layman does not know how a clock works, then he/she needs to learn how a mechanical clock works in the first place. This

Figure 3.9a A mechanical clock.

represents an amount of information or cost of entropy that he/she needs to learn. Secondly, after the layman has learned how a clock works, he/she has to figure out how to assemble it. This takes another set of information (i.e., equivalent to the amount of information provided by an instructional manual) to assemble the clock. So far, we have shown two major costs of entropy that the layman needs to assemble the clock. In practice, the cost is very excessive of which in practice the real cost of entropy needed is even higher! We further note that the information needed for the layman to assemble a clock is the dynamic-artificial-type information made by human, but not from the static-type information produced by Mother Nature's law of physics.

As for the follow-up question: How long will it take for the layman actually to assemble the clock. It will take him/her a lengthy time and effort (i.e., additional energy) to assemble it; my guess is it will take him/her a couple of days to a couple weeks.

Figure 3.9b   Parts of the clock.

Now, let us extend the preceding example further: What would be the cost of energy to actually create a mechanical clock shown in Figure 3.9(a), if we assume the mass of the clock is about one kilogram?

With reference to the Einstein's total energy equation, the magnitude of energy is equivalent to about 1000 atomic bombs dropped in Nagasaki, where we assume the creator can reverse it. Again the conversion from energy to mass would not give us a clock, but just a mass!

If we further assume that the creator has a foreseen objective to create such a mechanical clock shown in Figure 3.9(a), and he/she had decided to produce all the needed raw materials (e.g., iron ore, etc.) first, and let human to take care of the rest of tasks to eventually develop a clock. In this context, the creator needed an amount of information or a cost of entropy to producing those needed raw materials, in due process of converting energy into mass. Notice that this piece of information is achieved by the consequences (or chance of accidence) by the

Mother Nature's nuclei and chemical reaction. We notice that Mother Nature (or the creator) has all the energy to create all the needed raw materials, yet Mother Nature is not able to produce the man-made objects, such as clocks, computers, airplanes, and others. And it must be the reason that the creator (Mother Nature) left the rest of tasks for human to carry out the jobs in developing the clock!

Now, we assume we have all the needed raw materials, but we still have to process them before being used for building all the parts we need. And this is not the end of the story yet; we still need additional information to assemble the parts into a clock. If we add all of the aforementioned information made by human, it would sum-up to a great deal amount. Thus, by simply adding up the overall information needed (i.e., the ones contributed by Mother Nature and also the man-made), the creator would need a huge amount of entropy to create such a mechanical clock that he/she aiming at!

In the preceding example, we see that even though the creator (i.e., Mother Nature) has all the energy, but still cannot create the man-made objects! Although human is part of Mother Nature, yet there are things that human can do and Mother Nature cannot!

One more question may be asked: What would be the cost of energy and entropy to create an adult chimpanzee on this planet earth? And a follow-up question is that: What would be the chances for this creature to survive without the basic living skills in this hostile planet Earth?

This problem is more complicated than the mechanical clock example. Again if we assume the weight of the adult chimpanzee is about 40 kilograms, then the minimum energy for the conversion into mass is about 40,000 atom bombs dropped in Nagasaki. And the cost of entropy required for creating this chimpanzee (e.g., flesh, blood vessels, bones, nerves, DNA etc.) would be huge. We note that all those information required are primarily derived from the dynamic bio-evolutional processes, which cannot be accidentally created by Mother Nature's law of physics! Thus, it would cost an extremely large amount of entropy to instantly create this chimpanzee. Even though we assume that a creator may instantly be able to create this creature, the chance for this chimpanzee to eventually survive is extremely dim, without the basic living skills in this hostile planet Earth!

3.10   Remarks

In this chapter, we have shown the mythical relationship between science and information. Since every substance has a price tag or price tags of information which includes all the building blocks in our universe, one cannot simply ignore information when dealing with science. We have shown that there is a profound connection between information and entropy, in which information and entropy can be traded. Since entropy is a quantity well accepted in physics, it makes information a bit easier to connect with science.

Two of the most important pillars in modern physics must be the Einstein's relativity theory and the Schrödinger's quantum mechanics. We have shown that there exists a profound relationship between them by means of the uncertainty principle. In which we see that observation time window can be enlarged if the observer can keep up with light speed. In due of uncertainty relation with information, we have shown that every piece of information takes time and energy to transfer, to create, and to observe. In other words, every bit of information cannot be conveyed without a cost.

We have further shown that one of the most important aspects in digital communication is that a huge amount of information can be conveyed by light speed. Although the uncertainty principle was derived under the assumption of invasive investigation and it appears not to be violated, it does not mean one cannot exploit within the certainty limit. We have shown that wavefront reconstruction and SAR imaging can actually be obtained within the certainty region.

Since one cannot create something from nothing, we have shown that every substance in principle can be created with a huge amount of energy and a great deal of entropy! The question is that: Can we afford it?

Finally, I would stress that the cost of information is not meant to represent the information that one would like to get. It is the cost in terms of bit that one needs to pay for it. For instance, in a binary information source, the cost of each digital representation is "one" bit (i.e., equivalently to an amount of entropy). This is by no means that one willing to pay a bit (or the equivalent amount of entropy) that guarantees one will get the correct bit of information from the source that has provided. For example, it is the cost of an apple; it is by no means that if you paid for the cost of an apple that guarantees you will get an apple, you may get a roll of toilet paper which has the same cost. As a prominent astrophysicist in an interview saying that a book was threw into a black hole then somehow the black hole can give him back the equivalent amount of entropy. For which he thinks that he is able to get back the information of the book he threw in, but not knowing that he might get a different book has the same bits of information or an old computer that has the same cost of entropy.

References

1. C. E. Shannon and W. Weaver, *The Mathematical Theory of Communication*, University of Illinois Press, Urbana, IL, 1949.

2. F. T. S. Yu, *Optics and Information Theory*, Wiley-Interscience, New York, 1976.

3. F. W. Sears, *Thermodynamics, the Kinetic Theory of Gases, and Statistical Mechanics*, Addison-Wesley, Reading, MA, 1962.

4. L. Brillouin, *Science and Information Theory*, 2nd edition, Academic Press, New York, 1962.

5. E. Schrödinger, "An Undulatory Theory of the Mechanics of Atoms and Molecules," *Phys. Rev.*, vol. 28, no. 6, 1049 (1926).

6. W. Heisenberg, "Über den anschaulichen Inhalt der quantentheoretischen Kinematik und Mechanik," *Zeitschrift Für Physik*, vol. 43, no. 3–4, 172 (1927).

7. F. T. S. Yu, *Entropy and Information Optics*, Marcel Dekker, Inc., New York, 2000.

8. D. Gabor, "Communication Theory and Physics," *Phil. Mag.*, vol. 41, no. 7, 1161 (1950).

9. F. T. S. Yu, "Information Content of a Sound Spectrogram," *J Audio Eng. Soc.*, vol. 15, 407–413 (October 1967).

10. F. T. S. Yu, *Introduction to Diffraction, Information Processing and Holography*, MIT Press, Cambridge, MA, 1973.

11. D. Gabor, "A New Microscope Principle," *Nature*, vol. 161, 777 (1948).

12. L. J. Cultrona, E. N. Leith, L. J. Porcello, and W. E. Vivian, "On the Application of Coherent Optical Processing Techniques to Synthetic-Aperture Radar," *Proc. IEEE*, vol. 54, 1026 (1966).

13. A. Einstein, *Relativity, the Special and General Theory*, Crown Publishers, New York, 1961.

chapter four

Communication with Quantum Limited Subspace

4.1 *Motivation*

Most of the current information-transmission (IT) was constrained by the Heisenberg Uncertainty principle [1] as given by,

$$\Delta v \cdot \Delta t \geq 1 \qquad (4.1)$$

However, communication within the quantum limited condition [2], as given by,

$$\Delta v \cdot \Delta t < 1 \qquad (4.2)$$

has not been fully utilized yet. In this chapter, we will show that complex amplitude communication can be exploited within the quantum limited subspace (QLS). This tells us that there are ways to utilize communication space more innovatively, in which new communication ideas can be developed for practical application.

4.2 *Temporal (t > 0) Subspaces and Information*

As we had shown that [2], it is the time that ignited the creation of our temporal (t > 0) subspace and the created subspace cannot bring back the time that had been used for the creation. The essence of understanding our living subspace is that one cannot get something from nothing; there is always a price to pay, namely energy, time, and entropy (i.e., ΔE, Δt, and ΔS). Then it follows up a very fundamental question: Can we afford it?

With reference to the uncertainty relationship, we see that every piece of information needs an amount of energy ΔE (or equivalently the bandwidth Δv) and a section of time Δt to convey. And the **lowest** limit is given by,

$$\Delta v \cdot \Delta t = 1, \quad \Delta E \cdot \Delta t = h \qquad (4.3)$$

Uncertainty principle implies that **reliable information** is achievable if the particle is observed **under** the constraint of the uncertainty limit. And the uncertainty relationship also means that one **cannot** observe precise particle energy variation ΔE and its time Δt resolution as limited by the Planck's constant h. **Or equivalently,** one cannot determine the frequency resolution Δv and time Δt resolutions simultaneously of a particle under observation. In this context, we see that Heisenberg Uncertainty Principle represents an **IT limit**, as noted by Dennis Gabor where he named this limit as an **Information Cell** or a **Logan** [3], as illustrated in Figure 4.1, in which V_m and T are the wavelength and time limits of a time-signal.

Nevertheless, this wavelength-time plot can be subdivided into elementary **information cells**, as Dennis Gabor called **them Logons,** and each cell is bounded by the **lower bound of the uncertainty relationship** (i.e., $\Delta v \cdot \Delta t = 1$).

To confirm that **IT** is limited by the Uncertainty Principle, we show a set of speech spectrographic analyses, generated by a Bell-Labs sound spectrogram machine [4], as depicted in Figure 4.2. In which we see that time and spectral resolution (i.e., Δt and Δv) **cannot** be resolved simultaneously, with a single spectrographic displayed that is limited by the Uncertainty Principle (i.e., $\Delta v \cdot \Delta t = 1$).

We further stress that Uncertainty Principle shows that the relationship of an amount of energy ΔE (or bandwidth Δv) and interval of time Δt are needed for every bit of information to be transmitted. And the fact is that (ΔE, Δt) and (Δv, Δt) are real quantities. This is by no means

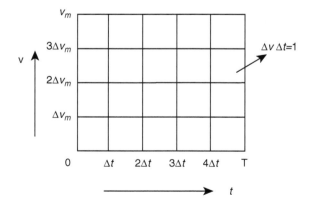

Figure 4.1 Gabor's information cells.

(a)

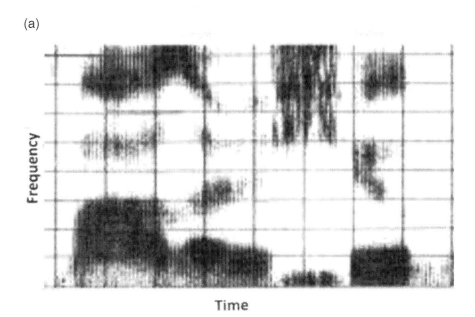

(b)

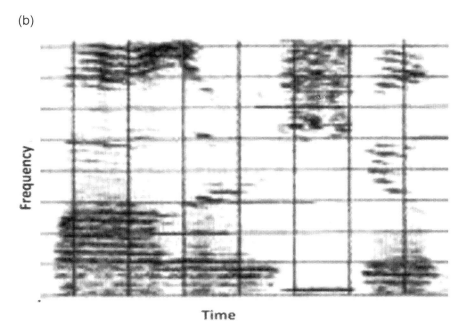

Figure 4.2   (a) Wide-band speech spectrogram. (b) Narrow-band speech spectrogram. Analyzing filter bandwidth Δv = 300 cycles/second. Analyzing filter bandwidth Δv = 100 cycles/second.

to imply that complex quantity observation (or communication) cannot be exploited within the certainty domain, as will be shown in our subsequent sections.

4.3 Quantum Unit

There are basically **two types of IT**: one is limited by **Uncertainty** Principle and the other is constrained within the **Certainty** relationship. And the **boundary between these IT regimes** is given by Eq. (4.3). For which we called this limit as a **Quantum Unit** [5]. We note that the shape of a quantum unit is not a critical issue, as long it is limited by 1 or h, respectively. For which we see that Δv and Δt can be simply **traded**. However, the important of IT under Heisenberg Uncertainty regime is that information is carried by means of **intensity** (i.e., amplitude square) variation, where **complex amplitude IT** has not been used. Then there is a **question** raised: Is it possible to transmit reliable information beyond the uncertainty limit (i.e., **within the certainty**) of which complex-amplitude information can be exploited? And the answer to this question is **yes**, as we shall show in the following.

There is, however, a major distinction between these two communication regimes, in which we see that most of the current IT fall within the Heisenberg's Uncertainty Communication regime. For instance, digital IT is one of the typical examples that are extensively used, which relies on the intensity variation to convey the digital information. Although evidence of exploiting complex-amplitude for IT within the quantum unit (i.e., certainty regime) has been successfully demonstrated such as wavefront reconstruction of Gabor [6] and Synthetics Aperture Radar Imaging [7], exploiting this regime for complex amplitude in communication is not recognized yet and its capability is not fully investigated.

4.4 Quantum Limited Subspace (QLS)

Since every temporal subspace in our universe can be described by time, where speed of light is the current limit, as given by $r = c \cdot \Delta t$, where r is the radius of the subspace. Then a Quantum Unit Subspace is limited by the Certainty relationship which can be represented as a **quantum limited-subspace QLS** [5] (or a **unit quantum subspace**) shown in Figure 4.3, in which we see a set of Rectangular QLS and Spherical QLS for (ΔE, Δt) and (Δv, Δt) as depicted in the Figure 4.3, respectively.

Since the carrier bandwidth Δv and time resolution Δt are exchangeable, we see that the size of the QLS enlarges as the carrier bandwidth

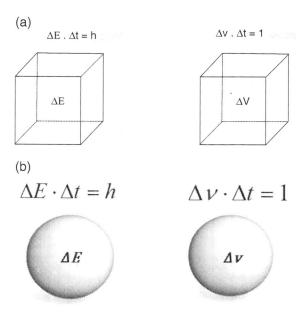

Figure 4.3 Quantum Limited Subspaces (QLS): (a) Cartesian Representation, side = cΔt; (b) Spherical Representation, radius r = c·Δt.

Δv reduces. In other words, narrower carrier bandwidth Δv has the advantage of having a larger QLS for complex-amplitude IT.

On the other hand, under the constraint of Heisenberg Uncertainty regime for IT, it is to **minimize** the size of QLS of which a narrower time resolution Δt can be used for rapid communication. As depicted in the Figure 4.4, we see that communication under the Certainty regime is to **maximize the QLS**, for which a larger (complex-amplitude) communication space can be utilized.

Let me stress again, one of the important aspects of the using QLS for IT is that complex-amplitude can be communicated **within** the subspace, for example, as applied to complex wavefront reconstruction (i.e., holographic recording) [6], where complex-amplitude information can be reproduced for observation and processing, for instance, such as complex-match filter synthesis of Vander Lugt [8]. On the other hand, IT under Uncertainty Constraint is to increase the digital transmission rate, where broader carrier bandwidth Δv would have the advantage for a faster IT rate.

Nevertheless, information can be transmitted either inside or outside QLS; for time-digital transmission, we can use a broader bandwidth Δv carrier which has the advantage of achieving a narrower Δt for rapid

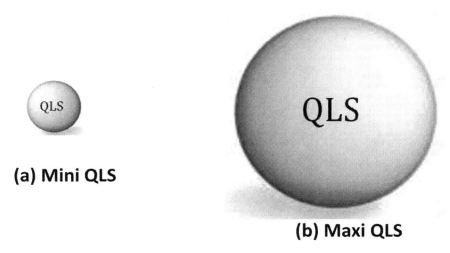

(a) Mini QLS

(b) Maxi QLS

Figure 4.4   (a) Mini QLS. (b) Maxi QLS.

transmission, but wider bandwidth Δv is more vulnerable for noise perturbation. As for frequency-digital transmission, we prefer a narrower bandwidth Δv at the expense of wider Δt, which has the advantage of lower noise perturbation but require longer Δt for transmission. Analog information-transmission has higher information content for time-digital or for frequency-digital transmission. Yet, the advantage of using digital-transmission is that it can be repeated and analog cannot. One of the major advantages of using digital-transmission is for noise immunity. Yet, there is a price to pay, which is by means of speed of light. In view of either time-digital or frequency-digital transmission, they are basically exploiting intensity (i.e., ΔE) for transmission which is limited by the Heisenberg Uncertainty Principle or equivalently information-transmission outside the QLS. For which we have shown that complex amplitude information can be exploited within QLS as follows.

4.5   *Examples of Information-Transmission within QLS*

To demonstrate that **complex-amplitude** IT can be implemented within the QLS, we show a conventional wavefront reconstruction result as depicted in Figure 4.5, in which a one-step rainbow hologram was made using a set of Argonne and a He-Ne lasers of about 6 **inches** coherence

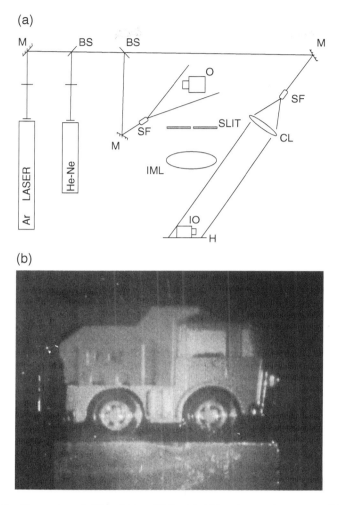

Figure 4.5 Example of IT within QLS: (a) Experimental set-up, (b) A 3-D color-hologram image.

length [9], which is about the **size** of the QLS for communication (i.e., observation).

Synthetic aperture radar (SAR) [7] is another example of IT within the QLS as shown in Figure 4.6. Since microwave antenna can be designed with a very narrow carrier bandwidth (i.e., Δv), its coherence length (i.e., $d = c \cdot \Delta t$) can be easily reached to hundreds of thousand feet. In other words, a very large QLS for complex-amplitude communication can be realized in practice. In which each returned pulse reflected from the targets carries the cross-range target-resolution at the same time brought in a complex wavefront from the reflected targets. And this is the complex wavefront from each reflected target per pulse carrier at a time to form a two-dimensional format for imaging, as depicted in Figure 4.6(b). In other words; the returned cross-range targets were imaged onto y-direction and the complex wavefront was sampled onto the azimuth x-direction to form a spatial encoded histogram. In view of the recorded format, we see that it is essentially a tilted one-dimensional hologram, in which the reflected wavefront was one-by-one sampled onto the y-direction of the recorded format, and it can be used for imaging as shown in Figure 4.6(c). Figure 4.6(d) shows an SAR image obtained by using the side-looking radar imaging technique, where the communication subspace (i.e., QLS) was well over 60,000 feet. In this example, once again we see that the narrower the carrier bandwidth Δv, the **larger the** QLS for IT, as we have seen that complex amplitude IT can be exploited for **remote sensing.**

4.6   Secure Information-Transmissions within QLS

In most practical IT, sender provides a carrier (e.g., electro-magnetic or acoustic wave) to transport the information to a receiver, for which an encrypted message is usually implemented for securely reason. Security IT is one of the major issues in current state of the art in communication. However, the more secure in IT the longer encrypted message is needed, which means that additional time and energy is needed for the encryption. Nonetheless, IT can also be done by using a different route of the message carrier. That is, instead of letting the sender provides a message carrier, it is the receiver provides the carrier to pick up the message from the sender. In this manner, a more secure IT can be delivered.

Let us assume an IT scenario as depicted in Figure 4.7, in which we see that the receiver send a pulse of carrier to pick up a message from a sender. Now we assume the sender using arrays of complex corner reflectors to attach a spatial message to the receiver. We see that a message can be imbedded with the reflected wavefront by the return pulses. If we further assume that the receiver divides the carrier into two pulses—one is sent to pick up the message and the other is retained for the coherent addition with the returned pulse wavefront—then we see that the embedded message can be unfolded using wavefront reconstruction idea, at the receiving end, provided that the traveling distance of the pulse carrier is within the QLS between the communicators. We note that one of the advantages of communicating within the QLS is that conferential information can be transmitted without using message encryption.

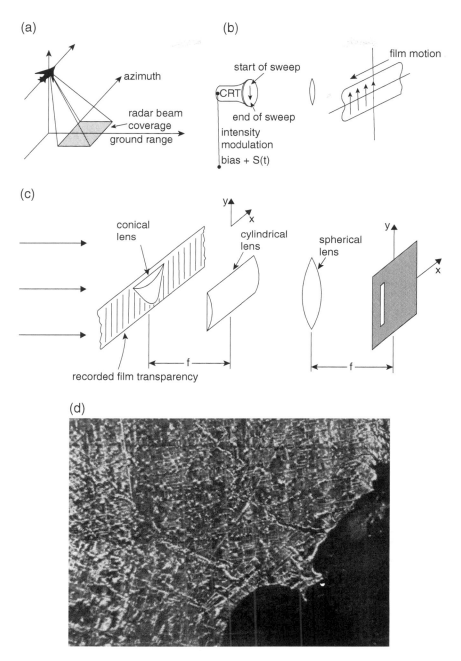

Figure 4.6 Side-looking radar imaging within QLS: (a) Side-looking radar; (b) Recording of the returned radar signals; (c) Optical processing for SAR imaging; (d) Sample of a SAR image.

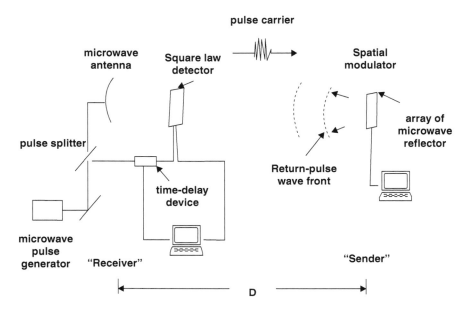

Figure 4.7   A hypothetical secure space communication configuration.

As for the coherence requirement, this is precisely the condition for IT within the QLS, in which complex amplitude wavefront can be communicated. Notice that, if the distance between the two communicators is **beyond** the coherence length of the pulse carrier (i.e., $\Delta d > c/\Delta v$), then it is **very unlikely or impossible** that the returned carrier wavefront can be **unfolded**. In which we see that IT within the QLS is protected naturally by the mutual coherence theory of wave propagation, since the intruder is **not** capable of producing a mutually coherent pulse carrier with respect to the returned pulse wavefront where the message is imbedded. Therefore, a confidential message can be securely transmitted within QLS, if the receiver sends a "signature carrier" to pick up the message. Of cause we had assumed that the sender has **a prior knowledge** of the incoming signature pulse carrier. Furthermore, additional protection for IT can also be implemented, such as using frequency hopping (i.e., spread spectrum) technique [10] that can be added in the pulse carriers. One can also employ message encryption to further strengthen the security in IT and so on.

We stress again, the major difference of the proposed secure communication format is that the receiver **provides the carrier** to pick up a message. As in contrast with the conventional IT, sender sends a carrier with a message. In this manner, information can be more securely transmitted **without** using message encryption.

4.7   Information-Transmission Using Phase Conjugations

Let us provide another example by which IT can be implemented using phase conjugation with a multi-mode fiber as shown in Figure 4.8. We all know that the spatial resolution transmission is depending upon the numbers of modes of the fiber; that is, the higher the modes, the higher the spatial resolution can be transmitted through a fiber. However, in practice, the transmitted output image is severely scrambled beyond recognition, although the information content of the image in principle remains. Although, in principle, the scrambled image can be unscrambled, if one launches the scrambled image through a perfectly identical fiber, using phase conjugation illumination, but the major impediment is that we simply cannot get a perfectly identical fiber in practice even though we have a perfect set up to do it.

In order to avoid having an identical fiber, we can simply interchange the positions of the sender and the receivers, as suggested in Figure 4.8. In which we assume that the receiver firstly launches a blank-image (i.e., without information content) at the input plane P_1 into the multi-mode fiber as depicted in the schematic diagram. Then a scrambled image as

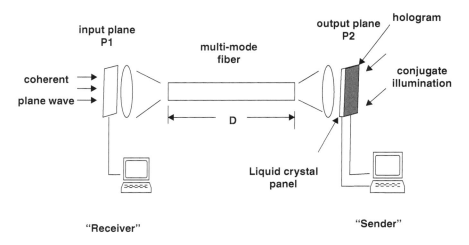

Figure 4.8   A fictitious information-transmission using phase conjugation.

from the input blank-image can be seen at the output plane P_2. If we assume that a hologram is made [11] at plane P_2, then by coherent conjugate illuminating the hologram, a conjugated wavefront of the scrambled blank-image (i.e., a reversal conjugated complex wavefront) can be launched back into the multi-mode fiber back, as shown in the figure. Then it is trivial to see that, a descrambled blank-image can be viewed at plane P_1. Let us insert an intensity image, as displayed onto the liquid crystal panel, behind the hologram at plane P_2 and using the same conjugate illumination. We would anticipate to seeing the inserted image at plane P_1, provided the distance of the fiber is well within the QLS (i.e., within the coherence requirement) as given by,

$$D \ll c/\Delta v \qquad (4.4)$$

where D is the length of the fiber, c is the speed of light, and Δv is the carrier bandwidth of the light source. Thus, in principle, information can be transmitted from the sender to the receiver using phase conjugation illumination. Therefore, we see that, as we used our creative ideas, I anticipate scores of innovative communication within the QLS will emerge, for which remains to be seen.

4.8   Relativistic Communication

Relativistic time at a moving subspace may not be the same as based on Einstein's special theory of relativity [12], as given by;

$$\Delta t' = \frac{\Delta t}{\sqrt{1 - v^2/c^2}} \qquad (4.5)$$

where $\Delta t'$ is the relativistic time window, compared with a standstill subspace; Δt is the time window of a standstill subspace; v is the velocity of a moving subspace; and c is the velocity of light. We see that time dilation $\Delta t'$ of the moving subspace, relative to the time window of the standstill subspace Δt, appears to be wider as velocity increases. For example, a 1-second time window Δt is equivalent 10-second relative time window $\Delta t'$. The means that 1-second time expenditure within the moving subspace is relative to about 10-second time expenditure within the standstill subspace.

Nonetheless, two of the most significant branches in modern physics must be Einstein's Relativistic physics and the Schrödinger's Quantum physics [13]. One is dealing with very large objects (e.g., universe) and the other is dealing with very small particles (e. g., atoms). Yet, there exists a profound relationship between them, which is by means of the Heisenberg's Uncertainty Principle [1] as given by,

$$\Delta v \cdot \Delta t \geq 1 \tag{4.6}$$

where Δv is the spectral bandwidth and Δt is the corresponding time resolution. This is the uncertainty relationship that represents a reliable bit of information transmission, in which we see that Δv and Δt can be simply traded; for example, the wider the carrier bandwidth Δv, the narrower the time resolution Δt for time-digital transmission, and the wider the Δt, the narrower the Δv for spectral-digital transmission. And the relationship between Einstein's physics and Schrödinger's mechanics is that, as viewed by the time dilation of Eq. (4.5), the relativistic uncertainty relationship can be written as,

$$\frac{\Delta v \cdot \Delta t}{\sqrt{1 - \left(\dfrac{v}{c}\right)^2}} \geq 1 \tag{4.7}$$

In which we see that the time-window $\Delta t'$ dilated somewhat relative to a moving subspace's time-window Δt of a standstill subspace (i.e., $\Delta t' > \Delta t$), as viewed from the moving subspace. This means that if one uses a time-digital signal format from a standstill subspace transmitting to a moving subspace, at the receiving end (i.e., moving subspace) time-resolution improves for faster time-digital transmission.

Since the spectral resolution is relatively improved (i.e., narrower) within a moving object as in contrast with a standstill subspace, one would use a digital-spectral transmission from a moving subspace to a standstill subspace, to gain the advantage of a finer spectral resolution in transmission. In principle, complex amplitude IT can also be exploited within the QLS as we have shown in preceding sections by using the relativistic narrower bandwidth from the moving subspace to the standstill subspace.

Basically there are two relativistic information-transmission strategies to select: one is using broader carrier bandwidth (i.e., Δv) and the other is using narrow carrier bandwidth. We have shown that broader carrier bandwidth is more susceptible to noise perturbation. Since the bandwidth of a carrier radiator is roughly proportional to its radiating frequency, the higher the carrier frequency, the higher the radiating bandwidth. Then there is a wide range of spectral carrier to select, as applied to relativistic information-transmission.

By taking the benefit of the relativistic time-window, one would use higher frequency for digital-time transmission from a moving plat- form to a stationary platform while using lower frequency carrier for digital-frequency transmission from a stationary platform to a moving platform.

4.9   Reliable Information-Transmission

One of the important aspects in communication must be the faithful information-transmission, in which reliable information can be reached with high certainty to the receiver. There are however two major orienta- tions in information-transmission; one was developed by Norbert Wiener [14,15] and the other was discovered by Claude Shannon [16]. Both Wiener and Shannon communication shared a common probabilistic basis; but there is a major distinction between them. The orientation of Wiener's communication is that if a signal (i.e., information) is corrupted by some physical means (e.g., noise or nonlinear distortion), it may be possible to recover the information from the corrupted one. It is for this purpose that Wiener develops the theories of correlation detection, optimum prediction, matched filtering, and so on. However, Shannon's communication carries a step ahead. He had shown that the information can be optimally transmitted provided that information is properly encoded. This means that the information to be transferred can be preprocessed before and after transmission. In the encoding process, he had shown that it is possible to counter the disturbances within the communication channel and then by properly decoding the encoded information, by which the received infor- mation can be optimally recovered. To do this, Shannon had developed the theories of information measure, channel capacity, coding processes, and so on. In other words, Shannon's information-transmission is efficient utilization of the communication channel. Thus we see that the basic distinction between Wiener and Shannon information-transmission is that Wiener communication assumes, in effect, that the signal in question can be processed after it has been corrupted by noise, while Shannon information- transmission suggests that information can be processed both before and after its transmission through the communication channel. Nevertheless, the main objective of these two branches of information-transmission is basically the same: a reliable information-transmission.

Since quantum entanglement researchers has used the fundamental principle of superposition for communication. Aside the fundamental principle is timeless (which does not exist within our temporal universe as will be seen in the next chapter), the underlying for information- transmission is illogical. Superposition principle in quantum mechanics means that multi-quantum states of an atomic particle exit simulta- neously and instantaneously, for which the quantum physicists felt they can utilize it for information-transmission. And it is the major motivation that quantum scientists want to exploit this fantastic phenomenon for communication [17] as well for quantum computing [18]. For example, as for quantum entanglement communication, they assume that the received information (e.g., binary form) from the sender should be "more equivocal" (i.e., ambiguous); the better the information transmission it is. As in contrast with the objective for reliable information-transmission, the sender wishes the message can be more reliably transmitted to the receiver. This is precisely the reason that sender uses a higher signal-to-noise carrier or lengthy redundancy signal and others so that unequivocal (i.e., unambiguous) information can be reached to the receiver. And it is not the objective for the sender to send an ambiguous signal to the receiver, to figure-out what message the sender intended to send. In effect of quantum entanglement communication we see that it is to retrieve ambiguous information from the sender; instead the sender should provide more reliable information. Quantum entanglement communication is designed for optimum extraction of information as Weiner type information retrieval, however, it is not for the purpose of efficient information-transmission of Shannon.

Aside the non-physical realizable of superposition principle within our universe, the fundamental aspect of using the superposition principle for the quantum entanglement communication is illogical, at least from the information-transmission standpoint. For example, for an information source, we want a larger "information content." It means that the more "uncertain" the information source is, the higher the information content provided by the source. On the other hand, as a receiver, the more "certain" the information received, the less amount of information content reaches the receiver. In other words, if the sender sends a signal of "1," the sender wants a higher certainty that the receiver will get it. However, it is the other way around in quantum communication; the sender wants the receiver to receive an ambiguous (i.e., unreliable) information for the receiver to guess what the sender had sent!

Finally I would stress that although information measure is equivalent to an amount of entropy, neither the information in bits (or other units) nor the equivalent amount of entropy represents the information content which has been frequently misinterpreted. The information measure is the "cost" in bits (or equivalent amount of entropy), but it does not represent the actual information. For example, if I am looking for a 2-bit book of [01], but there other 2-bit books of [11], [10] and [00] which are not the books that I am looking for. Thus, we see that the number of 10 bits book is equaled to $2^{10} = 1245$ copies of books that has the same bits of information but different contents! In other words, the book that you were looking for is still very uncertain even when you know the number of bits of your book (i.e., 10 bit).

4.10   Remarks

We have shown that every bit of information is limited by Δv and Δt, as we called it a **quantum unit**. Since every piece of temporal subspace is described by Δv and Δt, this unit can be translated as a quantum limited subspace (QLS), as imposed by the **certainty principle**. We have shown that IT can be carried out from inside and outside the QLS. That is equivalently to communication under the uncertainty constraint and under the certainty limit. In which we have shown that the **size** of a QLS is determined by **carrier bandwidth** Δv; that is, the smaller the carrier bandwidth, the larger the size of the QLS for complex amplitude communication. We have also noted that **wideband carrier** is more favorable for IT under uncertainty condition, while using **narrowband carrier** offers a larger QLS for complex amplitude IT. In other words, **wideband carrier** is more **suitable for** rapid IT, which is under the uncertainty regime. While using **narrowband carrier** provides a larger QLS for IT. We have also noted that wider carrier bandwidth offers higher transmission rate, but it is more vulnerable for noise perturbation. On the other hand, narrower carrier bandwidth provides a larger QLS, but it also slows down the transmission rate. In addition, we have shown that, without message encryption, **secure IT** can be communicated within the QLS. We have also added an example that spatial information can be, in principle, transmitted with a multi-mode fiber using phase conjugation illumination. We have also shown that relativistic communication between moving subspaces in principle is achievable. By taking the benefit of the relativistic uncertainty principle, viable information schemes for digital-time and for digital-frequency information transmission are achievable. Reliable information-transmission is that more dependable information can be received by the receiver. Using superposition principle for information-transmission is making the received information more ambiguous, which is not a reliable information transfer technique. Finally we predict that a new era of communication using QLS for innovative information-transmission is emerging. And it is anticipated that it will change the way we communicate (as well in observation and in computing) that we used to communicate, **forever!**

References

1. W. Heisenberg, "Über den anschaulichen Inhalt der quantentheoretischen Kinematik und Mechanik," *Zeitschrift für Physik*, vol. 43, 172 (1927).
2. F. T. S. Yu, "Time: The Enigma of Space," *Asian J. Phys.*, vol. 26, no. 3, 149–158 (2017).
3. D. Gabor, "Theory of Communication," *J. Inst. Elect. Eng.*, vol. 93, 429 (1946).

4. D. Dunn, T. S. Yu, and C. D. Chapman, "Some Theoretical and Experimental Analysis with the Sound Spectrograph", Communication Sciences Laboratory, Report 7, University of Michigan (August, 1966).

5. F. T. S. Yu, "Information Transmission with Quantum Limited Subspace," *Asian J. Phys.* vol. 27, 1–12 (2018).

6. D. Gabor, "A New Microscope Principle," *Nature*, vol. 161, 777 (1948).

7. L. J. Cultrona, E. N. Leith, L. J. Porcello, and W. E. Vivian, "On the Application of Coherent Optical Processing Techniques to Synthetic-Aperture Radar," *Proc. IEEE*, vol. 54, 1026 (1966).

8. A. Vander Lugt, "Signal Detection by Complex Spatial Filtering," *IEEE Trans. Inform. Theory*, IT-10, 139 (1964).

9. F. T. S. Yu, A. M. Tai, and H. Chen, "One-step Rainbow Holography: Recent Development and Application," *Opt. Eng.*, vol. 19, no. 5, 666–678 (1980).

10. A. R. Hunt, "Use of a Frequency-Hopping Radar for Imaging and Motion Detection through Walls," *IEEE Trans. Geosci. Rem. Sens.*, vol. 47, no. 5, 1402 (2009).

11. F. T. S. Yu, *Introduction to Diffraction, Information Processing and Holography*, Chapter 10, MIT Press, Cambridge, MA, 1973.

12. A. Einstein, *Relativity, the Special and General Theory*, Crown Publishers, New York, 1961.

13. E. Schrödinger, "Probability Relations between Separated Systems," *Mathematical Proc. Cambridge Phil. Soc.*, vol. 32, no. 3, 446–452 (1936).

14. N. Wiener, *Cybernetics*, MIT Press, Cambridge, MA, 1948.

15. N. Wiener, *Extrapolation, Interpolation, and Smoothing of Stationary Time Series*, MIT Press, Cambridge, MA, 1949.

16. C. E. Shannon and W. Weaver, *The Mathematical Theory of Communication*, University of Illinois Press, Urbana, IL, 1949.

17. K. Życzkowski, P. Horodecki, M. Horodecki, and R. Horodecki, "Dynamics of Quantum Entanglement," *Phys. Rev. A*, vol. 65, 012101 (2001).

18. T. D. Ladd, F. Jelezko, R. Laflamme, C. Nakamura, C. Monroe, and L. L. O'Brien, "Quantum Computers," *Nature*, vol. 464, 45–53 (March 2010).

chapter five

Schrödinger's Cat and His Timeless (t = 0) Quantum World

5.1 Introduction

One the most famous cats in science must be the Schrödinger's cat in quantum mechanics, in which the cat can be either alive or dead at the same time, unless we look into the Schrödinger's box. The life of Schrödinger's cat has been puzzling the quantum physicists for over eight decades as Schrödinger disclosed it in 1935. In this chapter, I will show that the paradox of the cat's life is primarily due to the underneath subspace in which the hypothetical subatomic model is submerged within a timeless empty subspace (i.e., $t = 0$). And this is the atomic model that all the particle physicists, quantum scientists, and engineers had been using for over a century, since Niles Bohr's proposed in 1913. However, the universe (our home) is a temporal space (i.e., $t > 0$) and it does not allow any timeless ($t = 0$) subspace in it. I will show that by submerging a subatomic model into a temporal subspace, instead of a timeless subspace, the situation is different. I will show that Schrödinger's cat can only either alive or dead, but not at the same time, regardless we look into or not look into the Schrödinger's box. Since the whole quantum space is timeless (i.e., $t = 0$), we will show that, the fundamental superposition principle fails to exist within our temporal space but only existed within a timeless virtual space. This is by no means of saying that timeless quantum space is a useless subspace. On the contrary, it has produces numerous numbers of useful solutions for practical application, as long it is not confronting the temporal and the causality condition (i.e., $t > 0$). In short, we have found that the hypothesis of Schrödinger's cat is not a physically realizable postulation and his quantum mechanics as well as his fundamental principle of superposition is timeless, which behaves like mathematics does.

One important aspect within our temporal universe (or time dependent universe) [1, 2] is that one cannot get something from nothing: There is always a price to pay. For example, every piece of temporal subspace (or every bit of information) takes energy and time to create. And the created subspace (or substance) cannot bring back the section of time that has expensed for its creation. Every temporal subspace cannot be a subspace of an absolute-empty subspace and any absolute empty space cannot have temporal subspace in it. Any science is proven within our temporal universe is physically real; otherwise it is fictitious unless it can be repeated by experiments. In other words, any analytical solution has to satisfy the fundamental boundary condition of our universe: dimensionality, temporal, and causality condition. Science is also a law of approximation and mathematics is an axiom of absolute certainty. Using exact math to evaluate inexact science cannot guarantee the solution exists within our temporal subspace. Science is also an axiom of logic, without logic science would be useless for practical application.

Added, all the fundamental sciences need constant revision. For example, science has evolved from Newtonian mechanics to Einstein's theory of relativity and to Schrödinger's quantum mechanics. And the beauty of the fundamental laws must be mathematical simplicity, so that complicated scientific logics and significances can be understood easily. And the advantages have been very useful for extending scientific researches and their applications. In fact, practically all the laws of science are point-singularity approximation, by which all laws were made to be broken and revised.

Nonetheless, practically all the particle sciences were developed from point-singularity model, which has no dimension and coordinates. But the atomic model [3] has been "unintentionally" embedding within an empty timeless (i.e., $t = 0$) subspace, as shown in Figure 5.1. Virtually it provides only the quantum state energy radiation $h\upsilon$, for which we have anticipated that limited new information can generate from this model, although Schrödinger has done an excellent job in building his quantum mechanics [4].

In view of Figure 5.1, we see that nucleus and electrons were shown by a dimensionless singularities representation. And we may not be aware that the model is in fact not a physically realizable model, since the submerged background represents a timeless empty subspace. However, timeless empty subspace cannot exist within our temporal universe! Although Bohr's atomic model have been used since the birth of his atom [3], but it has been mistakenly interpreted her background as an absolutely-empty timeless subspace. Strictly speaking, as a whole, it is not a physically correct model and the solution may not be used particularly as directly confronting the causality condition (i.e., $t > 0$). The reason is that the timeless subspace model (i.e., $t = 0$) cannot exist within our temporal space (i.e., $t > 0$).

On the other hand, any atomic model as presented in Figure 5.2 is a physically realizable model, at least satisfying the causality condition within our universe. In which we see that a Bohr atom is embedded within a temporal (time-dependent) subspace (e.g., our universe).

where ψ is the Schrödinger wave function, m is the mass, E is the energy, V is potential energy, and h is the Planck's constant. The description of Schrödinger equation shows that changes of a physical system over time, in which quantum effects takes place, such that wave–particle duality, are significant. However, the derivation of Schrödinger equation was based on point-singularity dimensionless atomic model submerged in a timeless (i.e., t = 0) empty space. And we have seen that there is a contrasting paradox, by which the model used for deriving the famous Schrödinger equation is physical realizably incorrect, since a time-dependent atomic structure was, inadvertently, submerged in an absolute-empty subspace. Of which the evaluated Schrödinger equation is also a timeless (i.e., t = 0) equation [5]. Let me stress that Schrödinger equation is a mathematical equation primarily developed to evaluate particle's quantum dynamics based on Bohr atomic model situated within an empty subspace. For which Schrödinger quantum mechanics is indeed a mathematic, any solution comes out from the Schrödinger equation cannot guarantee that exists within our temporal (i.e., t > 0) universe.

We note that the intention of using the timeless subspace Bohr model was inadvertently, since Bohr's atomic model has been success-fully accepted, in fact for over a century and we are still using this model. This may be the reason that causes us to overlook the basic assumption, of which a time-dependent (temporal) subspace should not be embedded in a timeless subspace, since they are mutually exclusive. Nevertheless, the essence of Schrödinger equation is to predict a particle probabilistic behavior, as a dynamic particle, by means of a wave func-tion. In other words, the outcome is not deterministic, but a distribution of possible outcomes. But a question is: Is Schrödinger equation a physically reliable equation to derive its wave equations? The answer is "no," as remained to be shown in the following:

Since the derivation of Schrödinger equation is based on **point-singularity** approximation, which is not a perfect assumption, yet it is an acceptable good approximation for this hypotheses. But it is the timeless subspace of the Bohr's atomic model embedded, which pro-duces timeless solutions (i.e., t = 0) that are not acceptable within the temporal (time-dependent) subspace. In other words, the solution as derived from Schrödinger equation is expected to be timeless since Schrödinger equation is a timeless equation. This is in contrast that the normally called Schrödinger equation is a time-independent equation, since timeless (i.e., t = 0) means time is not a variable while time-independent means time is an independent variable although Schrödin-ger equation has not have a time variable in it. Thus we see that Schrödinger's quantum mechanics is a timeless (i.e., t = 0) mechanics or timeless (i.e., with respect to the absolute-empty timeless subspace) mechanics, which does not exist within our temporal universe! As we have already shown, the adopted model of Figure 5.1 is not a physically realizable model, which should not have been used at all by Schrödinger. As we will show, his fundamental principle of super-position is the "core" of his quantum mechanics that is timeless (i.e., t = 0), which does not exist within our universe.

As quoted by the Richard P. Feynman [6], "He think he can safely say that **nobody** understands quantum mechanics. So do not take his lecture too seriously ... ". Yet, after we understood the flaw of Schrödinger's cat, which has haunted quantum physicists for over eight decades, we shall take a closer look at the paradox of the Schrödinger's cat. And at that moment, we may change our mind to saying that we have learned the inconsistency of Schrödinger's timeless (i.e., t = 0) quantum mechanics, as applied within our temporal universe (i.e., t > 0).

However, as I attempt to derive a wave dynamic where a particle is assumed to be situated within a temporal subspace, I am not sure that I will not be buried myself by complicated mathematical evaluation (e.g., I have not attempted to do it yet at the time being). But I anticipate that the new result would **not be** paying-off **at least for time being.** As I assume, my result will not have a better one than the Schrödinger equation that has **already been** developed at this time, since Bohr atomic model was oversimplified which has no dimension and provides little information for just quantum state radiation $h\upsilon$. Strictly speaking, the quantum state should be $h\Delta\upsilon$ since every physical radiator has a limited bandwidth. But I am sure my solution will comply with the causality condition (i.e., t > 0), if my quantum mechanical equation is developed using the temporal subspace model of Figure 5.2.

As has been done by using the Schrödinger equation **to** evaluate the particle wave function, one may need to reinterpret the solution to meet the causality constraint as imposed by our temporal universe. Other-wise, the evaluated solution **would not** be useful for practical applica-tion. In which we see that the **simultaneous existed multi-quantum states of an atomic particle** [7] is one of the typical examples that was derived from the classic Schrödinger superposition principle. And we can see that the simultaneous existent of multi-quantum states from atomic particle is "fictitious" and it would **not** happen. Firstly in view of the Bohr model, it is a point-singularity approximated model which has no dimension. Secondly within our temporal universe, time is distance and distance is time. Besides, every quantum state radiation (i.e., $h\Delta\upsilon$) is electro-magnetic in nature and has a bandwidth limitation that cannot simultaneously emit for all the quantum states. These are all the appar-ent physical reasons indicated to us that simultaneous existed multi-quantum states as promised by the classical superposition fails to exist within out temporal (i.e., t > 0) universe.

As we look back at the particle model embedded in an empty subspace of Figure 5.1 that Schrödinger based on developing his equation. In which we see that without such a simplified point-singularity model, viable solution may not be able to obtain even using tons of complicated mathematic manipulation. Although those assumptions alleviate (somewhat) the complexity for analysis, it also introduces **incomplete** and erroneous results that may **not** exist within our temporal universe. Thus by knowing Schrödinger's quantum mechanics, it is a timeless quantum computing machine, which was the consequence of using the assumed particle model within a timeless subspace. Since in practice timeless substance cannot exist within our temporal universe, we see that the flaw of Schrödinger cat, as well the whole Schrödinger's quantum world, is due to the assumption that the embedded subspace is absolute empty. In which we see that one cannot simply insert a timeless quantum machine into a time-dependent (i.e., t > 0) subspace.

5.4 Pauli Exclusive Principle and Particle Entanglement

Pauli exclusive principle [8] states that two identical particles with same quantum state cannot occupy the same quantum state simultaneously, unless these particles are existed with a different half-spin. While quantum entanglement [9] occurs when a pair of particles interacts in such a way that the quantum state of the particles cannot be independently described, even when the particles are separated by a large distance, a quantum state must be described by the pair of particles as a whole.

In view of Pauli exclusive principle, the entanglement between particles does exist, but the separation between the particles has to be limited, since the particles are situated within a time-dependent subspace (i.e., t > 0). Again, we see that the flaw of instantaneous entanglement comes from the assumption, when the exclusive principle was derived within the timeless (i.e., t = 0) subspace. In which we see again that temporal and timeless subspaces cannot coexist. In other words, time-dependent particles cannot coexist within a timeless subspace.

Nevertheless, instantaneous quantum entanglement [7] is one of the typical examples derived from the classic Schrödinger superposition principle. And we can see that the "instantaneous" (i.e., t = 0) entanglement between particles is "fictitious" and it would not happen within our temporal space. Quantum entanglement is depending upon the half-spin particle's quantum state energy of $h\Delta v/2$, which is apparently an electro-magnetic wave. For which the entanglement cannot go beyond the speed of light and cannot be instantaneous (i.e., t = 0), since time is distance and distance is time within our universe. In reality, the particle quantum entanglement distance is limited by its bandwidth Δv. As from all the physical evidences presented to us, instantaneous particle entanglement is false, only existed within an empty virtual space of mathematics, since super-position principle as will be shown is timeless.

Before we move away from the timeless issue, we would point out that practically all of the fundamental laws and principles in science, such as Paul's exclusive principle, Schrödinger's superposition principle, Einstein's energy equation, and others, are timeless principles and equations. Of which they were hypothesized "inadvertently" within an empty timeless environment.

5.5 Schrödinger's Cat

One of the most intriguing cats in quantum mechanics must the Schrö-dinger's cat, in which it has eluded the particle physicists and quantum scientists for decades. Let us start with the Schrödinger's box as shown in Figure 5.3; inside the box we have equipped a bottle of poison gas and a device (i.e., a hammer) to break the bottle, triggered by the decaying of a radio-active particle, to kill the cat. Since the box is assumed totally opaque of which we do not know that the cat will be killed or not, as imposed by the Schrödinger's superposition principle until we open the box.

With reference to the fundamental principle of superposition of quantum mechanics [4], the principle tells us that superposition holds for multi-quantum states in an atomic particle, of which the principle is the "core" of quantum mechanics. In other words, without the super-position principle, it won't have Schrödinger's quantum mechanics. In view of this principle, we see that the assumed two states radio-active particle inside the box can actually simultaneously coexist, with a cloud of probability (i.e., both one thing and the other existed at the same time).

Since the hypothetical radioactive particle has two possible quantum states (i.e., decay or non-decay) that existed at the same time, which is imposed by the virtue of superposition principle in quantum mechanics, this means that the cat can be simultaneously alive and dead at the same time, before we open the box.

But as soon as we open the box, the state of superposition of the radioactive particle collapses, without proof! For instant, we have found that, after the box is opened, the cat is either alive or dead, but not both. This paradox in quantum mechanics has been intriguing particle physi-cists and quantum scientists over eight decades, since the birth of Schrödinger's cat in 1935, as disclosed by Edwin Schrödinger who is as famous as Albert Einstein in modern physics.

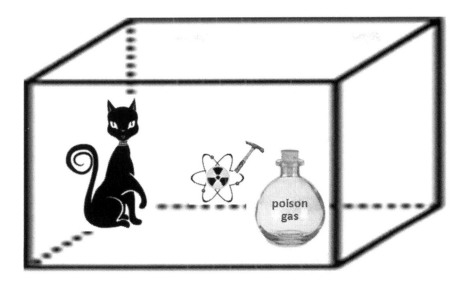

Figure 5.3 Inside the box we equipped a bottle of poison gas and a device (i.e., hammer) to break the bottle, triggered by the decaying of a radio-active particle, to kill the cat.

Let us momentarily accept fundamental principle holds such that superposition of a dual-quantum states radioactive particle exists within the box. This te'ls us that, the principle has created itself a Timeless (i.e., t = 0) Quantum Subspace or independent of time quantum space. However, timeless subspace cannot exist within our temporal universe. In which we see that, any solution (i.e., wave function) as obtained by Schrödinger equation contradicts the basic superposition principle, such that a timeless quantum subspace exists within our temporal (i.e., time-dependent) universe. This conjecture tells us that the hypothetical radioactive material cannot actually exist within the box, since both quantum states (i.e., decay or non-decay) cannot occur at the same time within a time-dependent subspace. We stress that time is distance and distance is time within a temporal subspace.

Nevertheless, it remains a question to be asked: Where is the source that produces the timeless radioactive particle? Or equivalently why Schrödinger's superposition principle is timeless (i.e., t = 0) for which the particle's quantum states exist simultaneously at the same time (i.e., t = 0)? A trivial answer is that it has to be coming from a timeless subspace where Bohr's atomic model embedded as shown in Figure 5.1. As we continue searching the root of paradox of the Schrödinger cat, we will provide an equivalent example to show that the paradox of the half-life cat is not a paradox.

5.6 *Paradox of Schrödinger's Cat*

Let us replace the binary radioactive particle with a flipping coin in the Schrödinger's box shown in Figure 5.4.

So as one flips a coin before it is landed, it is absolutely uncertainty that the coin will be either landed as a head or as a tail. Suppose we are able to "freeze" the flipping coin in the space at time t', then the flipping-coin is in a timeless-mode subspace at time t', which is equivalent to a two-state timeless particle freezes at time equates to t'. Then as soon as we let the flipping coin to continuingly flipping down at the same instance time $t = t'$, there should "no" lost time as with respect to the time of the coin itself, but "not" with respect to the time of the box. In other words, there is a section of time Δt that the box has gone by, in which there is a time difference between the coin's time and box's time. That is precisely why we cannot tell the cat will die or alive, as Schrödinger himself assumed his fundamental principle is correct. But as soon we open the box, we have to accept the physical consequence that the cat is either dead or alive, but not both. Then I guess Schrödinger created a logic to save his fundamental principle that superposition of the radioactive particle quantum states suddenly "collapses" as we open the box, without any physical proof. Otherwise the core of quantum mechanics fails to live up with the physical

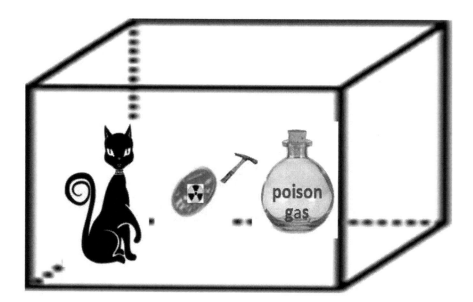

Figure 5.4 A flipping-coin analogy is substituted in the box for Schrödinger's Cat Paradox.

reality. Nevertheless, as we see it, the failure of the fundamental principle is due to the fact that a timeless flipping-coin "cannot be coexisted" within a time-dependent (i.e., t > 0) box.

We further note that it is possible to alleviate the timelessness of super-position, if we appropriately add the temporal constraint (i.e., t > 0) in deriving the Schrödinger's equation. We can change the timeless Schrödin-ger's equation to a time-dependent (i.e., t > 0) equation. Of which we will see that Schrödinger's wave functions of the dual-state radioactive particle can be shown as ψ_1 (t) and ψ_2 (t + Δt), respectively, where Δt represents a time delay between them. Since time is distance and distance within a temporal subspace, we see that the quantum states will not occur at the same time (i.e., t = 0). Furthermore, the degree of their mutual superposition states can be shown as a time ensemble of < ψ_1 (t) ψ_2*(t+Δt)>, respectively, where * denotes the complex conjugate. In which we see that a perfect degree of mutual superimposition occurs if and only if Δt = 0, which corresponds to the timeless (i.e., t = 0) quantum state of the radioactive particle.

Now let us go back to the half-life cat in the Schrödinger's box, where the radioactive particle is assumed within a timeless sub-box as shown in Figure 5.5. In which we see a timeless (i.e., t = 0) radio-active

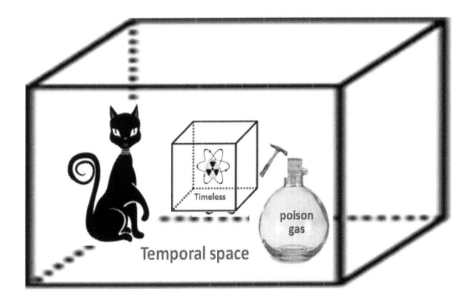

Figure 5.5 Schrödinger's Box with a timeless radioactive particle. Notice that, timeless radioactive particle cannot exist in a temporal (i.e., time-dependent) subspace.

particle is situated inside the time-dependent (i.e., t > 0) box, which is "not" a physically realizable postulation for the Schrödinger's Cat. The fact is that a timeless (t = 0) subspace cannot exist within a time-dependent (t > 0) space (i.e., the box). Thus, we have shown that again the paradox of Schrödinger's is not a paradox, since the postulated superposition is timeless and it is not a physically realizable principle within out temporal universe!

However, by replacing the timeless particle with a time-dependent (i.e., t > 0) particle shown in Figure 5.6, we see there is a match in time as a variable with respect to the box. Then Schrödinger's cat can only either be dead or not be dead but not at the same time, in which we see that there is nothing to do whether we open the box or not open the box to cause the fundamental principle to collapse. In other words, a dead cat or a live cat has already been determined before we open up the box. And the occurrence of the particle's quantum states is not simultaneously by means the fundamental principle of Schrödinger. In which we shown that superposition principle does not exist within our temporal space and it exists only within a timeless virtual subspace similar to what mathematics does.

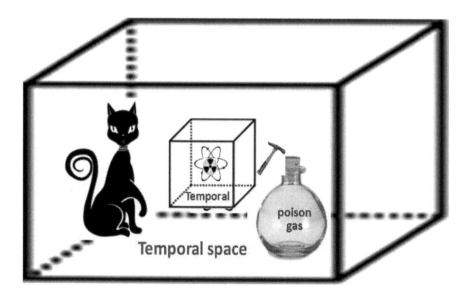

Figure 5.6 A time-dependent cat is in a temporal (time-depending) box. In which we see a temporal radioactive particle is introduced within the Schrödinger's temporal box.

At last, we have found the flaw of Schrödinger's cat, where Schrödinger was not supposed to introduce a timeless radioactive particle into the box. This vital mistake that he committed is apparently due to an atomic model in which subspace is assumed to be an absolute empty space as shown in Figure 5.1. In which we see that a timeless (i.e., t = 0) particle is wrongly inserted into a temporal (i.e., t > 0) box. I believe we have finally found the root of the paradox of Schrödinger's cat. For which we shall leave the cat behind with a story to tell; once upon a time there was a half-life cat ... !

5.7 Essence of a Subatomic Model

With high degree of certainty, most of the fundamental laws of science embraced the point-singularity approximation, which includes the atomic models embedded within a timeless subspace. As we look at any conventional atomic model, we might inadvertently assume that the background subspace is an absolutely empty space. And this is the consequence of Schrödinger's timeless quantum mechanics, since any physical atom (i.e., t > 0) cannot be situated within a timeless (i.e., t = 0) subspace, although singularity model works very well for scores of quantum mechanical application until the postulation of Schrödinger's cat emerged. Since the paradox of the half-life cat is the core of the fundamental principle, it has been arguing for over eight decades by Einstein, Bohr, Schrödinger, and many others since Schrödinger disclosed the postulation at a Copenhagen forum in 1935. This intrigues us to look at Schrödinger's equation which was developed on an empty (i.e., t = 0) subspace platform. In which we see that superposition principle collapses as soon as we open up the Schrödinger's box. This must be the apparent justification for Schrödinger to preserve the fate of his fundamental principle. Otherwise, his timeless fundamental principle cannot survive within our temporal universe (i.e., t > 0). In short, we see that the hypotheses of Schrödinger's cat are a fictitious postulation and we have proved that it does not have a viable physical solution, since any timeless radioactive particle cannot coexist in a temporal box and we have seen that Schrödinger has had it inadvertently introduced in the box.

5.8 Timeless Quantum World

Fundamental Principle of Quantum Mechanics tells us that superposition of a multi-quantum states particle holds if and only if within a quantum environment. By which it creates itself a Timeless Quantum Subspace, but timeless quantum subspaces cannot exist within our temporal universe. Then there is a question being asked: Are those timeless quantum subspaces can be utilized in our temporal universe? The answer is "no" and "yes."

The "no" part answer is that, if time component in application is an issue, such as applied to "instantaneous" quantum entanglement [9] and "simultaneous" quantum states computing [10], then the superposition principle as derived from the timeless Schrödinger equation would have problem, as applied within our temporal universe, since the superposition is timeless. For example, those instantaneous and simultaneous responses as promised by the fundamental principle do not exist within our temporal universe. And the postulation of Schrödinger's cat is not a physically realizable solution. In which we have shown that the burden of the cat's half-life can be liberated by using a temporal (i.e., t > 0) radioactive particle instead. In which we see that the paradox of Schrödinger's cat may never be discovered that it is not a paradox if we did not discover that Schrödinger's quantum mechanics is timeless.

Since Schrödinger equation is a timeless quantum computer, which is designed to solve a variety of particle's quantum dynamics, the solution as obtained from Schrödinger's equation is also timeless, which produces a non-realizable solution such as timeless (i.e., t = 0) superposition.

In which we see that, if one forces a timeless (i.e., t = 0) solution into a temporal (i.e., t > 0) subspace, one would anticipate paradox solution that does not exist within our temporal universe, such as the Schrödinger's half-live cat. This is equivalence of chasing a ghost of a timeless half-life cat in a temporal subspace. In which we have found that a timeless radioactive particle was inserted within the Schrödinger's box!

As to answer the "yes" part, if temporal aspect as applying a quantum mechanical solution is not an issue within our temporal space, then we have already seen scores of solutions as obtained from Schrödinger equation have been brought to use in practice, since the birth of Quantum Mechanics in 1933. This is similar to using mathematics (i.e., a timeless machine) to obtain solution for time-dependent application, which sometimes produces solution not physically realizable. In which we see that Schrödinger equation is a mathematics, which requires a time boundary condition (i.e., t > 0) to justify that its solution is physically realizable.

5.9 Quantum Mechanical Assessments

Schrödinger equation was developed on an absolute empty subspace platform, for which all the solutions are timeless (i.e., t = 0). Since the Fundamental Principle of Superposition was derived from the timeless Schrödinger equation, the corresponding quantum states' wave functions are also timeless with to respect the subspace that the particle is embedded in. Although wave function is time-dependent equation, it is with respect to the corresponding quantum state itself. This can be easily understood by an atomic model where the particle quantum states are represented by hv_n where n = 1, 2, ... N that represents the number of quantum states. In which we see that each n-th wave function is time dependent with respect to hv_n quantum state. And it is not with respect to the subspace that the atomic model is embedded in, which is an empty subspace. Since time-dependent wave function dictates the legitimacy of the superposition principle, the time dependency with respect to the particle's subspace is timeless, which is precisely the reason: the fundamental principle of superposition is timeless and the whole Schrödinger's quantum world is timeless (i.e., t = 0).

Since the whole quantum space is timeless, it cannot coexist within our temporal universe. In view, the logic of collapsing superposition principle as soon as we open up the Schrödinger's box is to satisfy the physical reality that the cat cannot alive and dead at the same time. Otherwise the fundamental principle of superposition has proved to be itself not existed within our temporal (i.e., t > 0) universe. It is apparent that Schrödinger's fundamental principle only exists within a timeless subspace. Personally, I believe this must be the reason for him to justify the fate of his fundamental principle; otherwise the principle is not able to survive. It must be Schrödinger himself who made the argument; otherwise the paradox of his haft-life cat has no physical foundation to debate by the world's top scientists over three-quarters of a century, since 1935.

Since quantum mechanics is a virtual quantum machine as mathematic does, we have found Schrödinger's machine is a timeless (or a virtual quantum) computer and it does not exist within our temporal universe. As we have seen that Schrödinger equation was derived within an empty subspace, and it is not a physically realizable model to use, since empty subspace and non-empty subspace are mutual exclusive. And we have seen that one plunges the timeless superposition principle within a temporal (i.e., t > 0) subspace and then antici-pates the timeless superposition to behave "timelessly" within a temporal subspace, which is physically impossible. In which we have shown that only mathematicians and quantum mechanists can do it, since quantum mechanics is mathematics.

But this is by no means to say that timeless quantum mechanics is useless, since it has been proven to us with scores of practical application but as long solutions that are not directly confronting with time-dependent or causality (i.e., t > 0) issue within our temporal universe. As quoted by late Richard Feynman [11] that "nobody understands part of quantum mechanics," yet we have found the part of nobody under-stands quantum mechanics which must be from the "timeless superposi-tion principle" that causes the inconsistency. And the root of timelessness quantum world is from the empty subspace that the atomic model was inadvertently anchored on. We are sure that this discovery would change our perception as applying the fundamental principle to quantum computing and to quantum entangle-communication. For which the "instantaneous and simultaneous" (i.e., t = 0 and concurrent) phenomena as promised by the fundamental principle do not exist within our temporal universe. Yet, important fallout from this discovery of the non-paradox of Schrödinger's cat encourages us looking for a new time-dependent quantum machine, similar to the one that Schrödinger has already paved the roadmap for us.

5.10 Remarks

In conclusion, I have shown that the atomic model that Schrödinger used must be anchored within an absolute-empty subspace. And it must be the underneath timeless subspace that caused the paradox of his half-life cat. The reason for overlooking the underneath timeless subspace must be due to the well-accepted Bohr's model that has been using for over a century, since the birth of Niles Bohr's atom in 1913 [3]. It has been very successfully used with excellent results for over a century. And it has never in our wildness dream that the underneath empty subspace causes the problem. In view of Schrödinger's time-dependent wave solutions, we have found the time dependency is with respect to the atomic particle itself but not with respect to the subspace the atomic model embedded in. In searching the root of the paradox of Schrödinger's cat, we found that a timeless radioactive particle should not have had introduced within a time-dependent (or temporal) Schrödinger's box. To alleviate the timeless radioactive particle issue, we have replaced a time-dependent (i.e., t > 0) radioactive particle for which we have shown that the paradox Schrödinger's cat is not a paradox after all. In short, we found the hypothesis of Schrödinger's cat is not a physically realizable postulation and his whole quantum world is timeless and behaves like mathematics does. Nonetheless, many of Schrodinger's timeless solutions are very useful until the implementation of fundamental principle that confronts with causality (i.e., t > 0) condition of our universe.

It may be interesting for me to note that Schrödinger's cat is observed by a group of non-communicating observers who take turns to look into the box. After each observation, the observer is required to close the box before for the next observer. Then how many times the fundamental principle needs to collapse and how many times the cat has to suffer the consequences for dying? My answer is that, if Schrödinger has had thought through the multi-life of his cat, we might not have had the puzzle that has been toying quantum physicists for over eight decades!

References

1. F. T. S. Yu, "Time: The Enigma of Space," *Asian J. Phys.*, vol. 26, no. 3, 149–158 (2017).

2. F. T. S. Yu, *Entropy and Information Optics: Connecting Information and Time*, 2nd Edition, CRC Press, Boca Raton, FL, 2017, pp. 171–176.

3. N. Bohr, "On the Constitution of Atoms and Molecules," *Philos. Mag.*, vol. 26, no. 1, 1–23 (1913).

4. E. Schrödinger, "Probability Relations between Separated Systems," *Math. Proc. Cambridge Phil. Soc.*, vol. 32, no. 3, 446–452 (1936).

5. L. Susskind and A. Friedman, *Quantum Mechanics*, Basic Books, New York, 2014, p. 119.

6. R. P. Feynman, R. B. Leigton, and M. Sands, *Feynman Lectures on Physics: Volume 3, Quantum Mechanics*, Addison-Wesley Publishing Company, Cambridge, MA, 1966.

7. C. H. Bennett, "Quantum Information and Computation," *Phys. Today*, vol. 48, no. 10, 24–30 (1995).

8. W. Pauli, "Über den Zusammenhang des Abschlusses der Elektronengruppen im Atom mit der Komplexstruktur der Spektren," *Zeitschrift Für Physik*, vol. 31, 765 (1925).

9. K. Życzkowski, P. Horodecki, M. Horodecki, and R. Horodecki, "Dynamics of Quantum Entanglement," *Phys. Rev. A*, vol. 65, 012101 (2001).

10. T. D. Ladd, F. Jelezko, R. Laflamme, C. Nakamura, C. Monroe, and J. L. O'Brien, "Quantum Computers," *Nature*, vol. 464, 45–53 (March 2010).

11. R. P. Feynman, R. B. Leighton, and M. Sands, *The Feynman Lectures on Physics*, Addison Wesley, Cambridge, MA, 1970.

chapter six

Science and Mathematical Duality

6.1 Introduction

Science is a law of approximation and mathematics is an axiom of absolute certainty. Using exact math to evaluate inexact science cannot guarantee that its solution exists within our temporal (i.e., t > 0) subspace. Science is also an axiom of logic. Without logic, science would be useless for practical application. Nonetheless, an ounce of good explanation is worth more than tons of calculation. Yet, the issue about burden of science is that it has to actually exist within our temporal space. In other words, any scientific (i.e., theoretical or mathematical) solution has to satisfy the boundary condition of our temporal space [3]. For example, dimensions and time are given by,

$$f(x, y, z; t), t > 0 \qquad (6.1)$$

where $f(x, y, z; t)$ is a three-dimensional space-time function in which time t is a forward variable and space and time coexist. In other words, any scientific solution which is not satisfied within the boundary conditions (i.e., temporal and spatial) of our temporal space is a virtual timeless solution that cannot be directly utilized within our temporal universe.

6.2 Subspaces

In this section we will introduce a few subspaces that have been known in practice: from an absolute-empty space, to mathematical virtual space, to time-independent Newtonian's space, and to Yu's temporal (i.e., t > 0) space. An absolute-empty space is that it has no substance and has no time. A virtual mathematical space is an absolute empty space that has no time but it has coordinates. In other words, virtual space has the coordinates but no physical substance in it. And it does not exist within our temporal (i.e., t > 0) space, since physical substance is time and time is physical substance. A Newtonian space [4] on the other hand is a space with substance in it. It has the coordinates, but treated time as an independent variable or dimension. And Newtonian space does not exist within our temporal space, since time and substance are mutually coexisted within our temporal universe. Temporal space [5] on the other hand is a time-dependent (i.e., $t > 0$) space, in which time is a forward variable at a constant speed that dictated by the speed of light. And space and times mutually coexist.

The important fact is that this temporal universe is our home and it is the physical space that we live in. Any physical constraint as imposed by our temporal subspace is a consequence by the creation of our universe, which was based on the laws of physics and the rule of time. In others words, we are the prisoners of our temporal space that includes all the physical sciences.

Physical fact is that any theoretical solution that deviates away from the constraints as imposed by our temporal space is not a physical solution within our universe. But this is by no means that the timeless absolute-empty space, virtual space, and Newtonian space are useless. On the contrary, they have paved the way in the history of mankind to learn new knowledge of science, for example as from absolute-empty space, virtual space, Newtonian space to temporal universe which are represented in Figure 6.1(a), (b), (c), and (d) respectively.

The differences among these subspaces are: absolute-empty is timeless and has no coordinates; virtual space is empty timeless and has dimensions; Newtonian space is time-independent and has coordinates; and Yu's temporal space is time-dependent (i.e., $t > 0$) in which substance and time coexist and time is a constant forward variable.

In view of these spaces, we see that an absolute-empty space has nothing in it and it is timeless; virtual space is kind of mathematical virtual-space, no physical substance, no time, or timeless (i.e., $t = 0$) and it has coordinates. Newtonian space is a time-independent space with

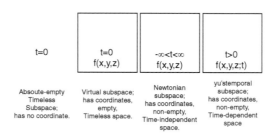

Figure 6.1 This figure shows an absolute-empty space (a), a virtual space (b), a Newtonian space (c), and a Yu's temporal space (d).

physical substance in it and it has coordinates. Since Newtonian space has treated time as an independent variable, time and space do not mutually coexist. While Yu's temporal space is a time-dependent variable space, in which time and physical substance mutually coexist, it is a time-dependent forward variable (i.e., t > 0) space.

Let me emphasize again, the distinction between Newtonian space and Yu's temporal space is that Newtonian space is a time-independent space in which time is an independent variable and it does not coexist with substance. While Yu's temporal space is that time is a dependent forward variable where time and substance coexist. And the speed of time is imposed by the speed of light, which is a dependent variable with space, as from the creation of our temporal universe [3,5].

There is however a difference in meaning between timeless space and time-independent space. Timeless space means that a space exists with no time (i.e., t = 0) where time is not a variable, while time-independent space means that a space exists at any time where time is an independent variable. Yu's temporal space means that a space coexists with time, where time is a constant forward variable and its velocity is already settled by the speed of light.

6.3   Subspace and Mathematics Duality

Our temporal universe is created by laws of physics and rule of time, which can be described by a spatial and time function as given by [3,5],

$$\nabla \cdot \mathbf{S} = f(x, y, z; t), t > 0 \qquad (6.2)$$

where ∇ is the divergent operator, $\cdot$ denotes the dot product, $\mathbf{S}$ an energy vector, (x, y, z) represents a spatial coordinate system, and time t is a constant-forward variable. Let me stress that; an equation is not just a piece of mathematical formula. It is a symbolic representation, a description, a language, a picture, or even a video. In which we see that this equation shows the creation of a subspace as time moves forward at a constant speed. This equation also shows that space and time coexist. In other words, any science proven within our temporal universe is physically real, otherwise it is fictitious. We stress that any solution as obtained by means of theoretical evaluation does not guarantee its existance within our temporal space, for instance, solution obtained from the set of Maxwell's Equations or by Schrodinger equation, unless their solutions satisfy our temporal subspace boundary condition, such as coordinates and temporal (or causality) condition. Of which we see that every physical science (i.e., physical subspace) coexists with time, and time is a constant forward variable (i.e., t > 0).

For any particle model embedded within any of those subspaces shown in Figure 6.1, the particle dynamic behavior is followed (or imposed) by the subspace that the particle model is embedded in. For example, if we mathematically plunge physically separated particles into an absolute-empty subspace, we will find out that all particles will be superimposing together, since the embedded subspace is timeless (i.e., $t = 0$), which has no coordinate and no distant (i.e., $d = 0$) between them. In other words, each particle can exist everywhere simultaneously within an absolute-empty space or timeless space. This is precisely what Schrödinger's fundamental principle of superposition is. But it only exists within a timeless subspace, since his quantum mechanics was built on a timeless empty-subspace platform [6].

Let me further stress that Schrödinger's timeless superposition principle should not be treated as a physical realizable science that can be implemented within our temporal subspace, for example, as applied to "instantaneous" (i.e., $t = 0$) particle entanglement and as well as applied to "simultaneous existence" quantum computing, which are fictitious solutions and not supported within our temporal subspace. However, timeless quantum mechanics can be redesigned to become temporal, if a quantum machine is built on a temporal (i.e., $t > 0$) subspace platform [7] as we will show in the subsequent sections.

Let us now use a subspace-system analysis to show their responses as from the respective system analysis as given in Figure 6.1, for example, from a timeless (i.e., $t = 0$) subspace-system, a temporal (i.e., $t > 0$) space-system, and a time-independent space-system, respectively, as a theoretical solution plunges (or input excitation) within one of the subspace-systems. Firstly, we let a uniform frequency distribution (i.e., Fourier transform) of Figure 6.2(a) as a pretended solution be used as an input excitation. Since Fourier solution is not a temporal domain or more precisely a time variable function, we cannot plunge the Fourier domain solution directly into any of the subspaces, unless it is converted (or inversely transformed into a temporal or a time domain solution), as depicted in Figure 6.2(c). Regardless of the non-physical realization issue, mathematicians and quantum mechanists can plunge any solution into an empty timeless space, since quantum mechanics has also been treated as mathematics. For system analysis, we see that the output response from an absolute-empty timeless is anticipated as given in Figure 6.2(e). In which we see that the output response occurs only at $t = 0$ the timeless domain. Since time is distance and distance is time within a timeless subspace, the output response ($t = 0$) exists all over the entire timeless subspace as illustrated topographically in Figure 6.2(f). In view of this system analysis, we see that the output response ($t = 0$) exists simultaneously all over the timeless subspace, which is precisely how Schrödinger's superposition principle behaves within a timeless

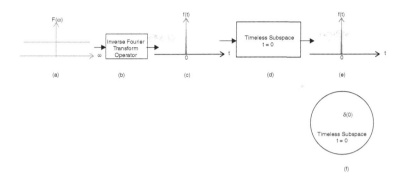

Figure 6.2   (a) An analytical frequency domain solution which is supposed to be used within our temporal subspace. Since Fourier function is not a time domain function, it needs to transform into a time domain equation before plunging into timeless subspace for test as shown in (c). By using this time function as an input excitation to a timeless (i.e., t = 0) subspace system is shown in (d), the corresponding output response is depicted in (e). And its output response is found everywhere within the timeless space domain as illustrated in (f).

(i.e., t = 0) subspace. Once again, timeless subspace (i.e., t = 0) cannot exist within our temporal subspace (i.e., t > 0), for which we see that the whole quantum world of Schrödinger is timeless which includes his fundamental principle.

Nevertheless, if we accept that the fundamental principle of super-position is a physically existed principle, then we will have a serious problem to reconcile with scientific reality. This is like searching a timeless angel within our temporal subspace! In which we have shown that Schrödinger's Fundamental Principle of Superposition does not hold within our temporal subspace. On the other hand, if we analytically force the timeless fundamental principle be applied within our temporal universe and pretending the superposition principle behaves "timelessly" within our universe, then unthinkable solution emerges which does not exist within our universe, for example, as the fictitious paradox of Schrodinger's cat, as well as all those "instanta-neous" information-transmission and multi-quantum states "simulta-neous" computing emerged, as promised by the superposition principle.

If we added a negative linear phase shift shown in Figure 6.3(b) with the Fourier distribution of Figure 6.3(a), the corresponding inversed transform in Figure 6.3(d) shows a time domain distribution of a delta function of $\delta(t - d/2)$ located at t = 0. Since the output response complies the causality (or temporal) condition (i.e., t > 0)

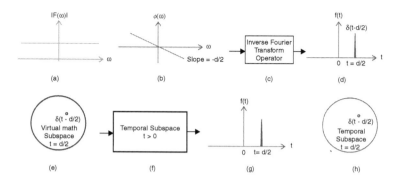

Figure 6.3 (a) and (b) A complex Fourier domain solution: The corresponding time-domain function is given in (d) which is launching into a temporal subspace system (f). (e) A time-topographical view at t = d/2 where the delta function located. (g) The output time response. (h) A time-topographic view where output response located, which shows the preservation of the input identity.

within our temporal subspace, it can be used or submerged within our temporal subspace. It is apparent that delta $\delta(t)$ represents a particle position in a timeless space at t =0, while $\delta(t-d/2)$ is the particle's location within a temporal (i.e., t > 0) subspace, since time is distance and distance is time within a temporal subspace as depicted from Figure 6.3(e) and (h), respectively. In which we see the output particle location is preserved. As in contrast within a timeless space, the output pulse exists "simultaneously" everywhere within a timeless subspace.

Let us provide a more convincing illustration shown in Figure 6.4, of which we assume a set of delta functions $\delta(t-t_1)$, $\delta(t-t_2)$, and $\delta(t-t_3)$, respectively, as shown in Figure 6.4(a). These represent a set of particles that is to be analytically plunging into a timeless subspace system in Figure 6.4(b) and the corresponding output response is shown in Figure 6.4(c). In which we see all the inputs (t-t$_1$), $\delta(t-t_2)$ and $\delta(t-t_3)$ are superimposingly converged at t = 0, where all the inputs lost their temporal identities within a timeless space. In other words within a timeless environment, it has no time, no coordinate, and no distance, by which all particles are superimposing at t = 0 and simultaneously existed everywhere within a timeless space. This is in fact the fundamental principle of superposition does, where multi-quantum states simultaneously existed. This illustration shows us that superposition principle only within a timeless space and it ceases to exist within in a temporal subspace where time is distance and distance is time as illustrated in a temporal-topographically diagram in Figure 6.4(d). While in Figure 6.4 (e), we see all particles are simultaneously superimposing at t = 0 and they also can be found everywhere within the entire timeless space, which is precisely what the fundamental principle does within a timeless space. Once again we have shown that all the "instantaneous" (i.e., t = 0) and "simultaneous" occurrences (i.e., d = 0) quantum states phenomena that superposition principle has promised only exist within a timeless subspace. For which one should not force the timeless fundamental principle to work within our temporal space; otherwise non-physical realizable solution will emerge.

Let us extend a previous complex Fourier domain example as depicted by Figure 6.5(a) and (b), in which we have added a positive linear phase factor with a uniform Fourier domain distribution. The corresponding inverse time domain solution is shown in Figure 6.5(d), where we see the time solution $\delta(t + d/2)$ occurs in the negative time-axis at t = -d/2. Since it is a negative time domain function, it is apparent that its temporal solution cannot be implemented within our temporal universe. Nevertheless, negative time response can be used within a Newtonian subspace, since Newtonian space treated time as an independent variable. Again, we note that Newtonian subspace cannot

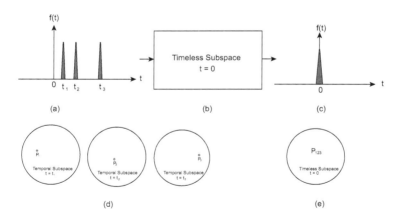

Figure 6.4 (a) An array of three-pulses (e.g., particles' locations) within a temporal subspace as shown in a temporal-topographical view in (d). As these particles plunge into a timeless subspace of (b), the output responses are super-imposing at t = 0 shown in (e) and the superimposed particles can be found all over the timeless domain. It is interesting to note that, within a timeless space, all things are in one and one is everywhere within the space.

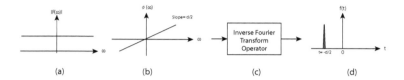

Figure 6.5 (a) and (b) A complex Fourier solution. (c) Inverse Fourier transform operator. (d) Its inversed-transform time domain function, which is located at t = -d/2; within a negative time domain.

be a subspace within our temporal space. In which we note that any temporal solution in part occurs in the negative time domain (i.e., t ≤ 0) is not a physical realizable solution to be used within our temporal (i.e., t > 0) universe. Otherwise, we can use the time independent Newtonian space for time travelling!

6.4   Causality and Temporal Universe

Every subspace within our temporal universe has a price tag; it takes an amount of energy ΔE and a section of time Δt to create it [3, 5]. Since time is distance and distance is time (i.e., d = c · Δt) within our temporal space, every particle P (or subspace) needs a quantity of (ΔE, Δt) to create and exist only at t > 0, as given by,

$$P(x, y.z; t) = P(t), t > 0 \tag{6.3}$$

where (x, y, z) denotes a three-dimensional spatial coordinate system and time t is a forward variable.

For simplicity in illustration, we assume a two-particle scenario for our consideration as given respectively by,

$$P_1(t - \Delta t) = \delta(t - \Delta t), t > 0 \tag{6.4}$$

and

$$P_2(t - \Delta t - d/2) = \delta(t - \Delta t - d/2), t > 0 \tag{6.5}$$

where $\delta(t - \Delta t)$ and $\delta(t - \Delta t - d/2)$ represents the particles' locations as shown in Figure 6.6,

In view of the temporal locations, we see that they satisfy the causality condition (i.e., t > 0) within our temporal universe; this means that the solution can be directly implemented within our temporal

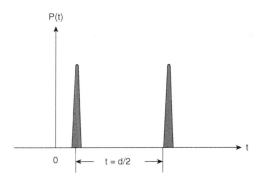

Figure 6.6 A two-particle scenario as denoted by two delta functions.

universe. Since time is distance and distance is time within our universe, the separation between particles can be shown as,

$$D = c \cdot (d/2) \qquad (6.6)$$

where c is the speed of light and $\Delta t = d/2$ is the "time" separation between the two particles.

Now if we submerged the two-particle scenario (i.e., time solution within our universe) of Figure 6.7(a) into our temporal universe, we will see there is a separation between the two particles within our temporal universe, since time is distance and distance is time within our universe.

On the other hand, if these particles are plunging into a timeless subspace, then particles lose their temporal identities which can only exist at t = 0 in a timeless space. In which the two particles are super-imposing together at t = 0 (i.e., a timeless space). Since time is distance and distance is time even within a timeless space (i.e., $d = c \cdot t$ and t = 0); the particles also lose their original positional identities and they exist everywhere within a timeless space as depicted in Figure 6.7(b), in which we see that within a timeless subspace it has no coordinate and no time. For which we stress again, all the "instantaneous" and "simul-taneous" existence quantum states are the traces of the Fundamental Principle of Superposition, but we have shown that the fundamental principle only exists within an empty timeless subspace and however the principle does not exist within our universe.

We shall now show a two more examples as we present a temporal solution into a lossless and a lossy temporal subspace system as depicted respectively in Figure 6.8(a) and (b). In which we see that,

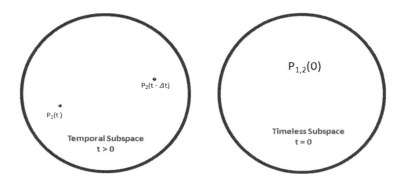

Figure 6.7 (a) Particles within a temporal subspace; (b) The superimposed particles within a timeless subspace, and the particles can be found simultaneously everywhere within a timeless space.

under lossless subspace scenario in Figure 6.8(a), the output response conserves the input energy and obeys the causality condition (i.e., t > 0). We see that the output responses are faithfully reproduced but with the same time delay as imposed by the causality constraint (i.e., t > 0) within our universe, where time is distance and distance is time. On the other hand, when we present this set temporal solution to a lossy temporal subspace system as shown in Figure 6.8(b), we found the output responses appear somewhat weaker and spread longer since the strengths of the inputs were absorbed somewhat within the lossy temporal universe. In view of the output responses, we see that overall output responses delay somewhat which is apparently due to the causality constraint within our universe (i.e., t > 0). It also tells us that, within our temporal universe, time and space (substance) coexist. In other words, causality condition (i.e., t > 0) is one of the most important fundamental constraints within our universe, of which our universe was created by means of a big explosion which was ignited by means of time, as a forward-dependent variable (i.e., t > 0) [3,5].

Another viable scenario is that, given a Fourier solution shown in Figure 6.9(a), its inverse time domain function is given in Figure 6.9(b), in which a part of its time domain exists in the negative time (t < 0) region that violates the causality (or temporal) condition within our temporal universe. In order to make the time solution physical realizable, we can introduce a linear phase shift in the Fourier domain as shown in Figure 6.10(a), so that it's inverse time function is entirely within the positive time domain as depicted in Figure 6.10(b). Thus we see that by properly delaying a time-domain solution, we can change

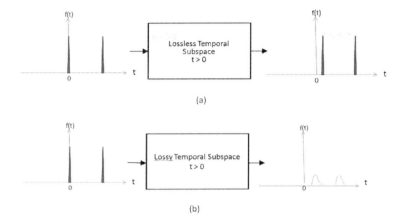

Figure 6.8 Output responses from a temporal subspace: (a) output responses from a lossless temporal subspace and (b) output responses from a lossy temporal subspace.

a non-realizable solution to a physical realizable one as illustrated in Figure 6.10(d). In which it tells us that, by simply changing the embedded subspace from a timeless subspace to a temporal subspace, it is possible to change from a timeless (i.e., t = 0) subspace model to become temporal (i.e., t > 0) physical model, which is a physical realizable model. In view of preceding illustration, a temporal quantum machine can be actually built on the top of a temporal subspace platform, instead on a timeless platform as Schrödinger did.

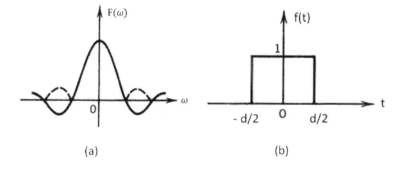

Figure 6.9 (a) A Fourier function. s a huge energy explosion

In view of preceding discussion, it tells us that, when an atomic model submerges into a subspace, its particle's quantum dynamic obeys the constraints as imposed by the subspace that the atomic model is embedded in. For example, when particles plunge into a timeless subspace, the particles dynamics loses its temporal and position identities (i.e., no coordinates, no time, and no distance) that obey the timelessness identities (i.e., instantaneous and simultaneous characteristics). Although it is not physically possible to implant a physical particle into an absolute-empty space from physical reality standpoint, but mathematically and quantum mechanically we could, since quantum mechanics is mathematics. But their solutions may not be temporal (i.e., $t > 0$), such as Schrödinger's quantum mechanics did. Similarly, if we submerge an atomic model into a Newtonian subspace, we see that the particles dynamic can behavior in the negative time domain (i.e., $t < 0$), which violates the temporal (or causality) condition of our temporal space. Even though Newtonian space is a time-independent space, it cannot be a subspace within our temporal universe. In view of all our illustrations, we see that the postulated Fundamental Principle of Superposition does not exist within our temporal universe by which all the "instantaneous and simultaneous" phenomena as the Fundamental Principle shown only exists within a timeless space, but does not exist within our universe.

Regardless of the mutual exclusive issues between timeless and temporal subspaces, some quantum scientists still believe that they can implant superposition principle within our universe. This is the reason that we would show what would happen when a multi-quantum states particle is implemented within a temporal space. For simplicity, we will simulate a two-quantum states particle which is plunging into an empty subspace as depicted in Figure 6.11. We further let two quantum states eigen values be $\exp[i(\omega_1 t)]$ and $\exp[i(\omega_2 t)]$, where ω represents the angular frequency of the quantum state. And the output response from

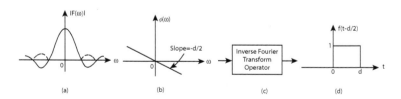

Figure 6.10 (a) and (b) A complex Fourier domain function. (c) Inverse Fourier transform operator. (d) The corresponding inverse-transform time domain function.

an empty space is given in Figure 6.11(c) that corresponds to a "timeless" (i.e., t = 0) dual-quantum state superposition, in which we assume energy is conserved. When this timeless simulated response (i.e., t = 0) is plunging into a temporal (i.e., t > 0) space as illustrated in Figure 6.11(d), its output response is shown approximately in Figure 6.11(e). We note that the output response occurs after at t > 0 and spreads all over after instantly t = 0, since time is distance and distance is time within a temporal space. In view of this simulated result, we learn that particle's quantum states firstly lost their personalities within a timeless space. After the empty space, we see the output response from the temporal space; it loses the original quantum states' properties. And the output response spreads out temporally (i.e., t > 0) and spatially, by virtue of energy conservation. In view of this output response of Figure 6.11(e), a three dimensional configuration of the response can be described as similar to Eq. (6.1);

$$\nabla \cdot \mathbf{S} = f(x, y, z; t), t > 0 \qquad (6.7)$$

where ∇ is the divergent operator, $\cdot$ denotes the dot product, $\mathbf{S}$ an energy vector, (x, y, z) represents a spatial coordinate system, and time t is a dependent-forward variable. Although it was a virtual simulated analysis, the result tells us that all the "instantaneous and simultaneous" existence quantum states as indicated by the superposition principle are not happening. For instance, this is precisely the simultaneous dual-quantum states radioactive particle that Schrödinger introduced in his

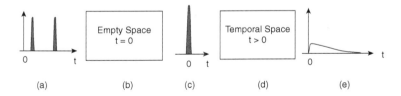

(a) (b) (c) (d) (e)

Figure 6.11 A hypothetical system as representing an empty subspace is embedded within a temporal space. (a) Input excitation, (b) Empty timeless system, (c) Output response from an empty space, (d) Temporal system, (e) The corresponding output response as from a temporal space. In which we see that the output response lost all the input temporal and spatial identities as given in Figure 6.11 (a).

box was a mistake, since he was anticipating that a timeless radioactive particle would behave "timelessly" within a temporal (i.e., t > 0) box.

6.5 *Flaws of point-singularity approximation.*

Practically all the laws of science are point-singularity approximated; otherwise those fundamental laws were being extremely difficult to be written in form of mathematics. This assumption also includes the simplistic atomic model used by particle and quantum physicists, for example, a typical Bohr atomic model has been frequently used as shown in Figure 6.12.

Although point-singularity approximation in science has the advantage for simplicity presentation, it is more difficult to describe for a multi-dimensional space, such as applied to a temporal subspace [i.e., f(x, y, z; t), t > 0] and others. For which we see that using simplistic subspace representation (i.e., set theory) has the advantage as compared if one tries using point-singularity approach. In view of the Bohr's atomic model, we see that the model shows no dimension, no coordinate, and no mass but provides only a piece of valuable (yet over

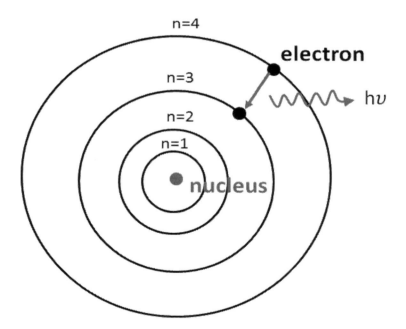

Figure 6.12 Bohr's atomic model.

simplified) information, the quantum state energy hυ (or radiation).
From the simplistic model, we would anticipate limited and in accurate
solutions come out from this model, no matter how complication the
solutions are. This limitation can be easily discerned as from information standpoint; less input information provided by the model produces
more limited viable output evaluation, unless the complexity of the
model extended. This is precisely the reason fictitious quantum sciences
emerged, since we have treated the unsupported postulation as
a physical real principle. By which some physicists even have the
notion that within atomic size regime, particle behaves strangely as
"Alice in the Wonder Land." One of the unsupported solution is
Schrödinger's fundamental principle of superposition, it has been treated the principle is an existed principle within our temporal universe for
over eight decades since the disclosure of the Schrödinger's cat in 1935.
Since then, there were countless fictitious analytically solutions emerged
from this timeless principle, but has not yet produced any viable result
to confirm their claims.

Let us show a system approach that superposition principle behaves
within an empty (i.e., timeless) subspace. And then follows up to show
how the temporal (i.e., t > 0) subspace reacted to a timeless superposition solution. Notice again within a temporal subspace, time and
subspace coexist; for which empty space is timeless and timeless space is
empty. Since we have treated point-singularity approximation atom, its
multi-quantum states can be described by means of wave equations as
given by;

$$\psi_n(x, t) = \exp\{i[kx - 2\pi\upsilon_n(t - \Delta t_n)]\}, n = 1, 2, \ldots N \qquad (6.8)$$

where k is the wave number, x is a spatial variable, υ is the quantum state
frequency, t is a forward time variable, and Δt_n is the section of time delay
with respect to nth quantum state. In which we see that each wave
equation is capable of emitting a package of wavelet (i.e., h$\Delta\upsilon_n$) at
different times, but not simultaneously. The point approximated multi-
quantum states model is submerged within an empty (or timeless) subspace as depicted in Figure 6.13, regardless of the legitimacy of substance
(i.e., atom) within an empty subspace issue, which mathematicians and
quantum physicists can do it; by virtue of energy conservation all the
emitting wavelets collapse at t = 0 within an empty subspace. And this is
precisely the timeless (or virtual mathematical) space that the fundamental principle of superposition is located, namely timeless (i.e., t = 0) space.

It is interesting for us to show a system analog using temporal
wavelets, as example representing two quantum states energy (i.e.,
h$\Delta\upsilon_n$, n = 1, 2) which are plunging into a virtual timeless subspace,

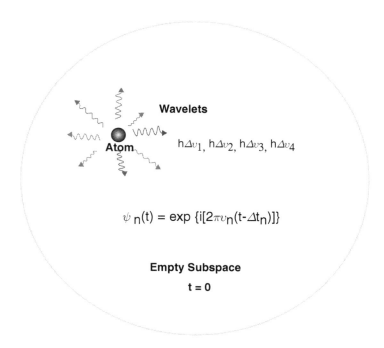

Wavelets

Atom

$h\Delta v_1$, $h\Delta v_2$, $h\Delta v_3$, $h\Delta v_4$

$\psi_n(t) = \exp\{i[2\pi v_n(t-\Delta t_n)]\}$

Empty Subspace

t = 0

Figure 6.13   A temporal atom is embedded within an empty subspace, although this is not a physical realizable model, since temporal and timeless do not coexist. But quantum mechanist can do it, since quantum mechanics is mathematics.

illustrated in Figure 6.14(a). In which we see that the two input wavelets lose their temporal personalities and collapse at t = 0, within an empty space. If we take this timeless response and plunging into a temporal (i.e., t > 0) space, we see that the response behaves temporally within a temporal subspace as shown in Figure 6.14(e), since time is distance and distance is time within our universe.

From this illustration, one sees that it is a serious mistake to implement a timeless superposition principle within a temporal space, even ignoring the physical realizable issue of a timeless space is a subspace within a temporal space. And it will be a more serious mistake as one is asking a timeless superposition to behave "timelessly" (i.e., t = 0) within our temporal (i.e., t > 0) universe. No wonder: "After you have learned quantum mechanics, you may not understand quantum mechanics," as quoted by later Richard Feynman. Now we might understand the part of quantum mechanics that we did not understand before; Schrödinger's fundamental principle of superposition is timeless and his quantum world is timeless.

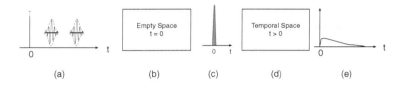

Figure 6.14   A system representation for a timeless subspace embedded within a temporal space. (a) A set of temporal wavelets plunge into a timeless space (b). (c) The output from the empty space. (d) The corresponding response within a temporal space, notice that the output response loses the original temporal personality of (a).

6.6   Dimensionless to Time-Space Transformations

Let us turn to the issue on dimensionality; in which our temporal subspace is dimensional (i.e., has spatial coordinates). Therefore, anything existed within our temporal space has to have coordinates; otherwise it cannot exist within our temporal subspace. Let us now take two of the most famous equations in science, namely the Einstein's energy [8] and Schrödinger's equation [9,10] as examples: one is responsible for the creation of our temporal universe and the other is used for the creation of the whole quantum world. Nevertheless, these two equations strictly speaking are timeless (t = 0). In other words, they are not time variable equations.

Let us start first with Einstein's energy equation, which is one of the most popular equations that more than three-quarters of the mankind know this equation. It is by means of its simplicity expression, but may not actually understand the significances and meaning by the majority. In terms of simplicity, Einstein's energy equation is given by,

$$\varepsilon = mc^2 \tag{6.9}$$

where m is the rest mass and c is the velocity of light.

And the other famous equation in quantum physics must be the Schrödinger equation as given by,

$$\frac{\partial^2 \psi}{\partial x^2} + \frac{8\pi^2 m}{h^2}(E - V)\psi = 0 \tag{6.10}$$

where Ψ is the Schrödinger wave function (or Eigen function), m is the mass, E is the energy, V is potential energy, and h is the Planck's constant.

In view of these equations we see that Schrödinger equation is far more complex than Einstein energy equation. Nevertheless, both equations are point-singularity approximations and dimensionless. And we also see that this set of equations is not temporal or time domain equations. Aside the dimensionless presentation, these equations are timeless (i.e., t = 0) equations. Because of their simplicity in mathematical representation, this set of equations has revolutionized the modern astrophysics and quantum physical world for over ten decades. Nonetheless, we are going to show some possible consequences when we use these equations, as analytical solutions, to directly plunging within our temporal (i.e., t > 0) universe.

Since our universe is a temporal (i.e., t > 0) expanding space, any analytical solution to be used directly within our universe has to be temporal and causal (i.e., t > 0). Therefore, by viewing this set of equations, we see that they are timeless (t = 0) equations, which cannot be implemented within our temporal universe, unless a time variable component can be appropriately introduced into these equations.

Let us now take Einstein's energy equation first as an example; in which we need to change the timeless equation to a temporal (or time-dependent) equation, so that it can be directly applied within our temporal universe. It is apparent that if we transform the Einstein energy equation into a partial differential form with respect to time, it is given by [3, 5],

$$\frac{\partial \varepsilon}{\partial t} = - c^2 \frac{\partial m}{\partial t} \qquad (6.11)$$

where the partial derivative of energy with respect to time $\partial\varepsilon(t)/\partial t$ is the rate of energy conversion, c is the speed of light, and the partial derivative of mass with respect to time $\partial m(t)/\partial t$ is the corresponding rate mass reduction. In which we have transformed the timeless Einstein equation into a time variable function. However, the partial differential for of Einstein equation is still a dimensionless singularity approximation, without showing any causality constraint (i.e., t > 0) on it. As mass is converting into energy process with respect to time, the equation can be represented with an energy diverging operation as given by,

$$\frac{\partial \varepsilon}{\partial t} = - c^2 \frac{\partial m}{\partial t} = \nabla \cdot S \qquad (6.12)$$

where $\nabla \cdot S$ is the divergent operator on a singularity energy vector S.

In view of this representation, we see that mass to diverging energy process shows how precisely our universe was created by means a huge energy explosion (i.e., big bomb theory). This expression shows from a point-dimensionless equation transformed into a three-dimensional expanding with time representation, in which time is a forward variable that dictated by the speed of light. And this is precisely how our universe which includes all the subspaces within our temporal universe can be described by following expression:

$$\nabla \cdot S = f(x, y, z; t), t > 0 \tag{6.13}$$

This equation also shows that every subspace (i.e., includes particle) within our universe is a temporal subspace and it is constrained by the causality condition (i.e., $t > 0$). In other words, every time response within our temporal universe cannot exist instantly (i.e., $t = 0$) but response at a later time (i.e., $t > 0$). And the equation also shows us that time and subspace mutually co-exist within our universe.

In view of all preceding example, we have shown that it is possible to transform a timeless equation to a time-dependent representation, which satisfies the causality condition within our temporal universe. We stress that an equation is not just a mathematical formula; an equation is a symbolic representation, it is also a description, a language, a picture or even a video, as may be seen from Eq. (6.12) and Eq. (6.13), in which we see that our universe was created by a huge energy explosion and its boundary is expanding at the speed of light, and the speed of time is settled by the light velocity.

We further show a set of well-known fundamental equations, such as Ampere, Faraday, Einstein, and wave equations as respectively given by;

$$\nabla \times E = -\frac{\partial B}{\partial t}, \ t > 0 \tag{6.14}$$

$$\nabla \times B = \mu_0 J + \mu_0 \varepsilon_0 \frac{\partial E}{\partial t}, \ t > 0 \tag{6.15}$$

$$\frac{\partial \varepsilon}{\partial t} = -c^2 \frac{\partial m}{\partial t}, \ t > 0 \tag{6.16}$$

$$\psi(x, t) = A \exp[i(kx - \omega t)], \ t > 0 \tag{6.17}$$

In view of these equations, we see that they are dimensionless temporal equations and have imposed by temporal or causality condition of $t > 0$, so that their solutions will be bounded within the causality condition of our universe.

Regardless it is non-physically realizable subspace within an empty space, what would be the consequences if we mathematically plunge the solutions as obtained from this set of equations into an empty subspace? Then the solutions will lost their temporal variability to timelessness state (i.e., $t = 0$). In other words, the solutions will converge to or superimposing at $t = 0$, and existed everywhere within the timeless space. In this it is precisely what superposition principle means with Schrödinger's quantum mechanics.

On the other hand, if we submerge their solutions in a temporal space, their temporal identities preserved and obeyed the forwarded time variation as imposed by the causality constraint (i.e., $t > 0$) within our universe. Thus, we see that it is the subspace of the atomic-model embedded that dictates the constraint of the temporal response. And the reason by imposing the causality constraint within those time-domain equations is just to be sure that their temporal solutions were guaranteed to be existed within our temporal (i.e., $t > 0$) universe.

6.7   Remarks

We have shown that science existed within our temporal universe is physical real; otherwise it is a virtual science as mathematics does. We have shown that there exists a duality between science and mathematics, in which an analytical solution can be shown; Does it exist within our temporal universe? We have also shown it is possible to reconfigure the solution to satisfy the causality condition within our temporal subspace. We have also shown that, as particles plunge into an empty subspace, all the particles are superimposed together at timeless (i.e., $t = 0$) space and they can be found anywhere within the timeless subspace. In which we have noted that Schrödinger's quantum mechanics is located within this timeless space. This tells us that his entire quantum world is timeless which includes his fundamental principle of superposition. Since the burden of mathematics is to prove first that there exists a solution for a mathematical postulation and then to find the solution. The burden of science is to prove that it exists within our temporal universe, and then experimentally to support it. Although Schrödinger's quantum mechanics is timeless, it has produced uncountable numbers of practical application, as long as her solutions are not directly confronting with the causality issue. However, it is his superposition principle that produces unsupported result within our temporal subspace, such as the paradox of his half-life cat. In short, timeless space is a virtual-space of absolute certainty at time zero (i.e., $t = 0$). Anything existed within a timeless subspace can be found instantly and simultaneously everywhere within the timeless space. Nevertheless, it is the timeless superposition principle that does not exist within our universe.

References

1. E. Schrödinger, "Probability Relations between Separated Systems," *Mathemat. Proc. Cambridge Phil. Soc.*, vol. 32, no. 3, 446–452 (1936).
2. E. Schrödinger, "Die Gegenwärtige Situation in Der Quantenmechanik (the Present Situation in Quantum Mechanics)," *Naturwissenschaften*, vol. 23, no. 48, 807–812 (1935).
3. F. T. S. Yu, "Time: The Enigma of Space," *Asian J. Phys.*, vol. 26, no. 3, 149–158 (2017).
4. O. Belkind, "Newton's Conceptual Argument for Absolute Space," *Int. Stud. Phil. Sci.*, vol. 21, no. 3, 271–293 (2007).
5. F. T. S. Yu, *Entropy and Information Optics: Connecting Information and Time*, 2nd ed., Boca Raton, FL, CRC Press, 2017, 171–176.
6. F. T. S. Yu, "The Fate of Schrodinger's Cat," *Asian J. Phys.*, vol. 28, no. 1, 63–70 (2019).
7. F. T. S. Yu, "A Temporal Quantum Mechanics," *Asian J. Phys.*, vol. 28, no. 1, 193-201 (2019).
8. A. Einstein, *Relativity, the Special and General Theory*, Crown Publishers, New York, 1961.
9. L. D. Landau and E. M. Lifshitz, *Quantum Mechanics*, Pergamon Press, Oxford, 1958, 50–128.
10. M. Bartrusiok and V. A. Rubakov, *Introduction to the Theory of the Early Universe: Hot Big Bang Theory*, World Scientific Publishing Co., Princeton, NJ, 2011.

chapter seven

Temporal (t > 0) Quantum Mechanics

7.1 Schrödinger's Timeless (t = 0) Quantum Mechanics

Now, let us look back at the Schrödinger's equation [1] as given by,

$$\frac{\partial^2 \psi}{\partial x^2} + \frac{8\pi^2 m}{h^2} (E - V)\psi = 0 \tag{7.1}$$

where Ψ is the Schrödinger wave function (or Eigen function), m is the mass, E is the energy, V is potential energy, and h is the Planck's constant. In view of this Schrödinger equation, we see that it is a dimensionless and timeless (i.e., t = 0) equation as anticipated, since his quantum machine was built by a singularity Bohr atom anchored on the top of an empty subspace. Since it is not a temporal equation, it cannot be directly implemented within our universe which is a temporal (i.e., t > 0) space [2,3]. Added Schrödinger equation does not represent the relative-position of the particle's quantum states, since atomic particles have always been treated as a point-singularity object where the sizes and positions of the subatomic particles were neglected.

Since Bohr's atomic model is self-contained, it seems nothing to do with the subspace that the atomic model is embedded in. This is similar to one walks within a continental air craft; it seems nothing to do with our motion (or time) as with respect to the motion of our planet. However, it makes a huge difference as the atom model submerges into a timeless or a temporal subspace. If an atomic particle is submerged into a timeless subspace, their sub-atomic particles' quantum states and locations (as had been show in the proceeding chapter) will be superimposing together and also located everywhere within a timeless subspace, as within a virtual mathematical space does. Since quantum mechanics is itself mathematics, similar to the set of Maxwell equations, the solution as obtained from the Schrödinger equation does not guarantee the existence within our temporal universe. Firstly the solution has to be a temporal (or time variable) function and secondly it has to comply with the causality condition (i.e., t >0) of the temporal universe, where its solution (i.e., wave equation) is supposed to submerging or applying in. However, when an atom model is submerged into a time-dependent or a temporal subspace, the non-coordinate subatomic particles become dimensional, since temporal subspace has coordinates. And also time is distance and distance is time within a temporal subspace [2, 3].

Nevertheless, the objective for a quantum scientist is to calculate the particle-wave duality dynamics, as derived from the Schrödinger's equation, in which its wave dynamic solution is given by [4],

$$\psi(x, t) = A \exp[i(kx - \omega t)] \tag{7.2}$$

where A is an arbitrary constant, $k = 2\pi/\lambda$, λ is the wavelength, $\omega = 2\pi\upsilon$, and υ is the radiant frequency. In view of this equation, we see that the particle-wave duality has essentially two variables (x; t) as expected from the Bohr's atomic model [5]; one represents a traveling wave in a positive spatial x direction and the other shows a wave moving in opposite direction in time t variable. In which we see that the variables (x; t) are profoundly connected to the particle's quantum state energy (i.e., $\Delta E = h\Delta\upsilon$), but shows no sign of the quantum state position, because Bohr's atomic model is point-singularity approximated. However, as quantum physics moves to distance or to sub-subatomic scale application, the relative quantum state position and the subatomic scale dimension cannot be ignored, for example such as applied to instant particle entanglement communication [6] and as applied to simultaneous quantum-state computing [7].

However, we shall treat each quantum state particle-wave duality as a point radiator; then their wave dynamic presentations can be written in terms of eigen state as given by,

$$\psi_n[\omega_n(t - \Delta t_n)] = A \exp.\{-i[\omega_n(t - \Delta t_n)]\} \tag{7.3}$$

where n = 1, 2, 3, ... , Nth quantum state. Thus we see that all the quantum states of a multi-quantum states particle under consideration are "not" superimposing together, since time is distance and distance is time within a temporal subspace. As in contrast with superposition principle has promised, all the quantum states exist simultaneously. We note that Bohr's atom model [5] has been regarded as a point-singularity approximation for over a century. But in reality, every subatomic particle has mass and dimension; no matter how small they were, they cannot be totally ignored. Thus we see that a perfect superposition of particle's quantum states is impossible.

7.2   Searching for a Temporal (t > 0) Quantum Machine

Let us now look at the possibility of developing a temporal (i.e., t > 0) quantum machine similar to the one that Schrödinger had built. The answer is yes; but it has two approaches to be considered. As from a rigorous approach, we have to build a new quantum machine on the top of a temporal atomic model as depicted in Figure 7.1(a), instead of using the model shown in Figure 7.1(b) that Schrödinger had used where a Bohr's atom is inadvertently submerged within an empty timeless subspace.

With reference to Schrödinger quantum mechanics [8, 9], it would take us some time to build a temporal quantum machine, if I would? Since it is not my present objective, I have opted to leave it behind for the time being. Nevertheless, from an information theorist standpoint, I would first start to replace an empty platform with temporal subspace where Schrödinger's quantum machine had anchored on. This is precisely what I have done in restructuring the existing quantum machine to become temporal; a Bohr's atomic model is built on the top of a temporal platform shown in Figure 7.1(a).

Since the Bohr's atom model is self-contained, unfortunately plunged into a wrong timeless platform shown in Figure 7.1(b). In view of the atom model, we see that Schrödinger's wave equation is time variable (i.e., time-dependent) equation, of which its eigen value is given by,

$$\psi(\omega t) = \exp[-i(\omega t)] \tag{7.4}$$

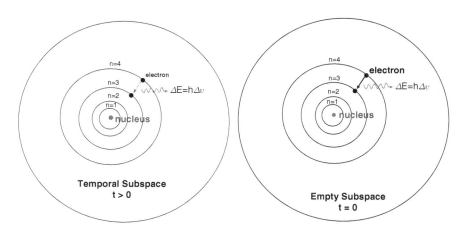

Figure 7.1   Two subatomic models. (a) A Bohr atom is embedded within a temporal subspace. (b) A Bohr atom is embedded in an empty timeless subspace. h is the Planck's constant and v is the radiation frequency. Notice, I have changed $E = hv$ to $\Delta E = h\Delta v$, where Δv is the bandwidth of quantum state energy.

where $\omega = 2\pi\upsilon$ and is the respective angular frequency and t is a time variable as obtained from a particle-dynamic wave duality similar to Bohr's model as given by Eq. (7.3), and υ is the quantum state radiation frequency as emitted from the quantum-state energy $\Delta E = h\Delta\upsilon$. As we look back at Bohr's atomic model, we see that the model provided us only two pieces of information, namely quantum state energy (i.e., $\Delta E = h\Delta\upsilon$) and mass (m). But it has no coordinate, since atomic particle's size has been assuming so small and treated it as a point-dimensionless radiator. This is precisely that Schrödinger's wave function shows no coordinate or position in it. However, as quantum entanglement is moving to distance application or as applied to simultaneous quantum states computing, the particle's position cannot be ignored. In which we see that, time-dependency wave function is only with respect to the Bohr's atom model itself (or self-contained), but not with respect to the empty subspace that Bohr's atomic model is embedded in as shown in Figure 7.1(b). On the other hand, if a Bohr's atom is submerged within a temporal subspace as shown in Figure 7.1(a), time-dependent wave function is complied by temporal or causality condition (i.e., t > 0) within the temporal subspace. In which we see that the atomic model in Figure 7.1(a) is a physical realizable model, which exists within our temporal universe.

By ignoring the non-physical realizably model of Figure 7.1(b), we would like to show how the quantum-state energy $\Delta E = h\Delta\upsilon$ behaves within an empty subspace. We note that the speed of quantum-state radiation is instant (i.e., t =0) and the radiation energy (i.e., $\Delta E = h\Delta\upsilon$) can be simultaneously existed everywhere within a timeless space, which is precisely what fundamental principle of superposition has promised, but unfortunately the principle only exists within a virtual timeless space. And this must be the reason that many physicists think that Schrödinger's Cat is a scientific paradox, which is exactly the same reason that many quantum scientists use the "instance and simultaneous existed" quantum states (i.e., superposition principle) for quantum entanglement [6] and quantum computing [7]. But, unfortunately the fundamental principle of superposition in quantum mechanics does not exist within our temporal universe.

Nonetheless as in view of Figure 7.1(a) and (b), we see that υ is the radiated frequency of the quantum-state energy. Therefore, every bit of quantum computing is limited by the Heisenberg Uncertainty Principle [10] as given by,

$$\Delta E \cdot \Delta t \geq h, \quad \Delta\upsilon \cdot \Delta t \geq 1 \qquad (7.5)$$

We note that if the half spin of Pauli's exclusive principle [11] is taken in consideration for particle entanglement; its Heisenberg Uncertainty Principle can be written as,

$$\Delta E \cdot \Delta t \geq h/2, \ \Delta v \cdot \Delta t \geq 1/2 \qquad (7.6)$$

where Δv is bandwidth of the quantum-state energy ΔE. Although the wider the quantum state bandwidth Δv, the narrower the time resolution Δt, it also shortens the particle entangle distance or coherence length [12] as given by,

$$d \leq \Delta t \cdot c = c/\Delta v \qquad (7.7)$$

where c is the velocity of light.

In view of uncertainty principle of Eq. (7.5), it tells us that every bit of quantum computing or every quantum entanglement is not unlimited within our temporal universe. In which we see that it is limited by the inherent quantum-state bandwidth Δv, as in contrast with the fundamental principle of superposition that has shown instant response (i.e., t = 0), simultaneous existence everywhere without any physical limit. Since time resolution Δt is limited by Δv, the speed of quantum entanglement (as well for quantum computing) will not go beyond the speed of light, which was assuming by the superposition principle. In view of the atomic model shown in Figure 7.1, information-transmission either for quantum entanglement or simultaneous quantum computing has to carrier by the particle's quantum state energy $\Delta E = h\Delta v$. For which every bit of information transmitted is limited by bandwidth Δv of the particle's quantum state carrier, which is imposed by the Heisenberg uncertainty principle.

Nonetheless, the implicit description of the wave function represents a particle wave dynamic behavior as described by eigen-value of exp[-i(ωt)], which is a time-dependent function but with respect to Bohr's atom itself (i.e., self-temporal). Yet as an atomic model submerged (mathematically) in an empty subspace, the atomic model behaves as a whole with respect to the subspace that the atom model is submerging in, which is timeless. This is similar as one walks within a trans-continental aircraft; walking is with respect within the aircraft but not with respect to the rotation of our planet earth. One may ask if the aircraft is embedded within an empty space (although it is not a physical realizable assumption), what would be your walking inside the aircraft with respect to the empty space that the aircraft is embedded in? The fact is that wave function cannot specify the particles position within the atom, since that subatomic model has no dimension. Nevertheless, the purpose of wave function is to describe a particle-wave duality dynamics by means of a wave representation. As a matter of fact, any physical model submerges into an empty space is not a physically realizable model, since substance and absolute emptiness are mutually exclusive. But, mathematicians and quantum mechanists can make it happen by anchoring a temporal atom on a virtual space, since quantum mechanics is a computational instrument as mathematics does!

In view of Bohr's atomic model as depicted in Figure 7.1(b), it was and still is a very viable model for particle physicists and quantum mechanists. And it has been used over a century and has never been encountering any major problem until emerges of non-supported postulations such as the paradox of Schrödinger's cat [13] and as well as all problematic "instant" quantum entanglement and "simultaneous existed" quantum states for computing, as promised by the superposition principle. Of which we have shown that timeless superposition principle cannot exist within our temporal (i.e., t > 0) universe. Nonetheless, the consequence of timeless superposition principle is the solution obtained from a quantum machine that is anchored within an empty platform. By which the fundamental principle of superposition has been treating as physically real, since the emerging of Schrödinger's equation in 1935.

On the other hand, if one submerges a Bohr's atom into a temporal subspace such as the one shown in Figure 7.1(a), then multi-quantum states relative positions have to specify, since time is distance and distance is time within a temporal space. For which particle's quantum states positions can be presented by an incremental component Δt_n within the eigen-value of the wave function (i.e., $t - \Delta t_n$). Therefore, we see that separation between n^{th} quantum state with respect to the first quantum state can be written as,

$$d_n = c \cdot \Delta t_n, n = 2, 3, 4, \ldots, N \qquad (7.8)$$

where c is velocity of light and Δt_n is the time separation between quantum state number 1 with respect to nth quantum state. And the respective eigen value of the respective wave function can be written as,

$$\{\psi_1(\omega_1 t), \psi_2[\omega_2(t - \Delta t_2)], \ldots \psi_N[\omega_N(t - \Delta t_N)]\} \qquad (7.9)$$

where $\omega_n = 2\pi v_n$ is the angular frequency of the nth quantum state, n = 1,2 3, ... N and Δt_n is the time separation with respect to reference quantum state n=1, which can be translated into distances as shown by,

$$d_n = c \cdot \Delta t_n \qquad (7.10)$$

Figure 7.2 shows two hypothetical diagrams, in which Figure 7.2(a) represents a set of particle's quantum states distributed within a temporal subspace, where their positions can be physically identified by $d_n = c \cdot \Delta t_n$ since time is distance within a temporal (i.e., t >0) space. While in Figure 7.2 (b) we see a set of output wave functions where their positions cannot be specified, since they are situated in a timeless (i.e., t = 0) environment. This is precisely where Schrödinger's quantum world located instantly at t = 0. I would further emphasize that timeless quantum subspace cannot exist within our temporal space, period. That is precisely the reason that paradox of Schrödinger's cat is not a paradox [13].

In view of particle-wave duality as described in Eq. (7.8), a mutual cross-quantum state coherence (or simply cross-coherence) function between quantum states number 1 with respect to nth quantum state can be defined as giving by,

$$\Gamma_{1n}(\Delta t_n) = \ <\psi_1(\omega t)\psi_n^*(\omega_n(t - \Delta t_n)> \quad (7.11)$$

where the < > represents a time ensemble operation and * denotes as a complex conjugate. This equation essentially can be written as,

$$\Gamma_{1n}(\Delta t) = \lim_{T \to \infty} \frac{1}{T} \int_0^T \psi_1(t)\psi_n^*(t - \Delta tn)dt \quad (7.12)$$

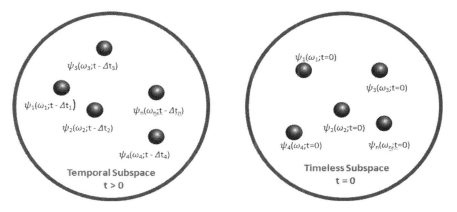

Figure 7.2 Distributed hypothetical eigen-value wave functions within a temporal and a timeless subspace, respectively. (a) A scenario where eigen-values occur at different time within a temporal subspace. (b) A scenario where all the eigen-values occur at the same time t = 0 within a timeless virtual subspace.

And their degree of cross-coherence between n^{th} quantum state with respect to quantum states number 1 can also be shown as,

$$\gamma_{1n}\ (\Delta t) = \frac{\Gamma_{1n}\ (\Delta t)}{\left[\Gamma_{11}\ (0)\Gamma_{nn}\ (0)\right]^{1/2}} \qquad (7.13)$$

In which we see that a perfect mutual degree of cross-coherence between two quantum states occurs at $\Delta t_n = 0$, which is identical to the case under timeless condition (i.e., $t = 0$), where Schrödinger's fundamental principle of superposition is located at $t = 0$. However if $t = 0$ and $\Delta t_n = 0$ (i.e., $d_n = 0$), we see that particles' multi-quantum states collapsed at the same time $t = 0$. If we accept point-singularity approximation, we found the fundamental principle of superposition holds if and only if they are situated within a timeless quantum space at $t = 0$. In other words, the locations of all the particles' wave functions (or quantum states) "occur" at $t = 0$ within a timeless environment. But we know that timeless space is a virtual mathematical space which cannot exist within our temporal universe.

Since each quantum state emits different wavelength of radiation (i.e., v_n), a high degree of mutual cross-coherence between different quantum states is very unlikely. Nevertheless, if we split-up an emitted quantum state beam into two paths, then an auto-coherence function between these two beams can be written as,

$$\Gamma_{nn}\ (\Delta t) = \lim_{T \to \infty} \frac{1}{T} \int_0^T \psi_n(t)\psi_n^*(t - \Delta t)dt \qquad (7.14)$$

where $n = 1, 2, 3, \ldots N$, $\Psi_n(t)$ is the wave function of nth quantum state, Δt is the time separation between the two beams, and $*$ denotes the complex conjugate. The degree of the auto-coherence is given by,

$$\gamma_{nn}\ (\Delta t) = \frac{\Gamma_{nn}\ (\Delta t)}{\Gamma_{nn}\ (0)} \qquad (7.15)$$

In which we see that, we see a perfect degree of auto-coherence occurs at $\Delta t = 0$, where there is no difference in path length between the two beams. In other words, two beams travel at an equal distance.

In view of the timeless quantum subspace shown in Figure 7.2(b), we see that all the wave functions superimposing together and simultaneously distributed all over the entire timeless subspace. This is exactly what superposition principle means; multi-quantum states can simultaneously exist anywhere within the subspace. Therefore, under point-singularity approximation, fundamental principle of superposition exists only within an empty timeless space (i.e., t = 0), but it is the fundament principle that cannot be used within a temporal space (i.e., t > 0), since timeless and temporal are mutually excluded. In fact, Figure 7.2(b) is virtual mathematical model which does not exist in reality, since empty and non-empty (i.e., timeless and temporal) are mutually exclusive. And this is precisely what Schrödinger's quantum mechanics had done to us, although his quantum machine has serving us with scores of viable application.

Although it is not physical realizable but quantum mechanical possible from virtual mathematical stand point, we assume a set of various quantum states wave functions are plunging into an empty timeless subspace shown in Figure 7.3. Since it is an empty subspace, we see that velocity of the quantum state radiation is infinitely large

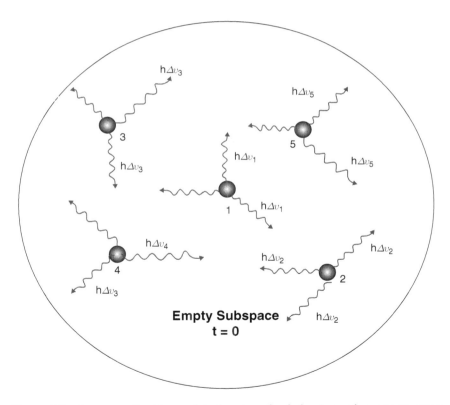

Figure 7.3   A non-realizable model showing the behaviors of quantum state radiations within an empty timeless subspace.

(i.e., instant, t =0) and simultaneous as anticipated by means of the fundamental principle of superposition. This illustration is to convince the readers that Schrodinger's fundamental principle, the core of his quantum mechanics only exists within a timeless virtual space since quantum mechanics is treated as mathematics. Therefore, it is a vital mistake to apply the core of fundamental principle within a temporal subspace, since the timeless superposition does not exist within our universe. In which we see that all those "simultaneous" and "instantaneous" information-transmission as promised by the fundamental principle are fictitious and virtual.

7.3   Pauli Exclusive Principle and Particle Entanglement

Pauli exclusive principle [11] states that two identical particles with same quantum state cannot occupy at the same quantum state simultaneously, unless these particles exist with a different half-spin. Quantum entanglement occurs when a pair of particles interacts in such a way that the quantum state of the particles cannot be independently described; even when the particles are separated by a large distance, a quantum state must be described by the pair of particles as a whole. In view of Paul's principle, we assert that, the atomic model used for his discovery has no coordinate (or position). This is a reasonable assumption since atomic size particles are very small, for which singularity approximation is appropriately used in most of the times.

However, as quantum communication move to distance application (or simultaneous quantum states computing), then the temporal and causality (i.e., t > 0) issue cannot be ignored. In which we found that separation between entangled particles becomes a problem, since time is distance and distance is time within our temporal universe. For example, as depicted in Figure 7.4(b) when two particles are entangling within a timeless quantum space. Since within an empty space it has no time, no coordinate, and no distance, we see entangling particles is instant (i.e., t = 0) and can be entangled everywhere within a timeless space. Notice that in reality, physical particle (i.e., temporal particle) cannot exist within an empty subspace. It is only quantum mechanists who can implant a temporal atomic model into an empty subspace, in which quantum mechanics behaves as virtual mathematics does.

On the other hand, if we submerge a pair entangling particles within a temporal environment as shown in Figure 7.3(a), the situation is rather different; we see that the wave functions are physically separated by d =c ·Δt, since time is distance. In which we see that particle cannot entangle instantly within temporal universe.

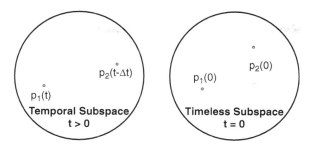

Figure 7.4   Particle quantum entanglement: (a) Particle entanglement view within a temporal (i.e., t > 0) subspace. (b) Particle entanglement view within a timeless subspace (i.e., t = 0).

In view of the Pauli's half-spin exclusive quantum state as expressed by the uncertainty principle in Eq. (7.6), we see that it half-spin's entangled distance is limited by the radiation bandwidth Δv. Similarly, a mutual entangling function between the two half-spin entangled particles can be written as,

$$\Gamma(\Delta t) = \lim_{T \to \infty} \frac{1}{T} \int_0^T \psi_1(t)\psi_2^*(t - \Delta t)dt \qquad (7.16)$$

where Ψ_1 and Ψ_2 are the two entangling wave functions, * denotes the complex conjugate, t is a time variable as obtained from a particle-dynamic wave duality, and Δt represents the time separation between entangling particles such as $\Delta t = d/c$, where d is the physical separation between the two entangled particles.

Again, we see that a perfect entanglement occurs at $\Delta t = 0$ (i.e., the two particles are superposing together as referred to Eq.(7.14), we have,

$$\Gamma(0) = \lim_{T \to \infty} \frac{1}{T} \int_0^T \psi_1(t)\psi_2^*(t)dt \qquad (7.17)$$

This equation is only truth under point-singularity approximation of the particles. In view of the mutual entangling function, we see that the strength of entanglement decreases rapidly as separation (i.e., Δt) increases, since within temporal subspace, the separation between the entangled particles is d = c Δt. In other words within our universe, one cannot get something from nothing and there is always a price to pay, namely energy and time (i.e., ΔE and Δt). In which quantum entanglement cannot be the exception and we have also shown that "instant" entanglement (i.e., t = 0) does not exist within our temporal universe (t > 0). Once again we stress that fundamental principle of supposition in quantum mechanics only exists within a timeless virtual space (i.e., t = 0) and it does not exist within our temporal universe (i.e., t > 0).

In view of all preceding arguments, we see that a temporal subatomic model is needed for all the subatomic quantum analyses; otherwise virtual fictitious solution emerges without substantiated verification. We have shown that using timeless (i.e., t =0) particle entanglement as pretending existed within our universe is physically impossible, since timeless and temporal are mutually exclusive. And the distance of mutual entanglement is limited by the degree of mutual entanglement between two particles as given by,

$$\gamma_{12}(\Delta t) = \frac{\Gamma_{12}(\Delta t)}{\left[\Gamma_{11}(0)\Gamma_{22}(0)\right]^{1/2}} \tag{7.18}$$

Since Schrödinger's quantum mechanics was built with a dimensionless Bohr atomic model which has only two pieces of information—quantum state energy $h\Delta v$ and mass m of an electron—any solution comes out from this atomic model cannot exceed the information that has been provided by the this atomic model (i.e., $h\Delta v$, m). Of which we see that Schrödinger wave equation {i.e., $\Psi(x, t) = A \exp[i(kx - \omega t)]$} is an electromagnetic wave equation. Any quantum state radiation from this wave equation cannot beyond the speed of light and its radiating energy is limited by the quantum state bandwidth $\Delta\omega$ (or Δv). In view of the quantum state bandwidth, quantum entanglement is limited by mutual entangle length as given by Eq. (7.16) which shows that the entanglement is not instant (i.e., t = 0) as superposition principle has promised. In which we see that quantum entanglement is possible within our universe, but it is limited by speed of light. And the particle entangling distance is limited by the mutual entanglement length as imposed by the Pauli's exclusive half-spin principle as given by the following uncertainty relation such as,

$$\Delta E \cdot \Delta t \geq h/2, \quad \Delta v \cdot \Delta t \geq 1/2 \tag{7.19}$$

In which we see that the narrower the half-spin quantum state bandwidth Δv, the longer the mutual entanglement length, that is d = c/(2 Δv) where c is the speed of light.

7.4   Virtual and Temporal (t > 0) Duality

The most important aspect of an analytical science is that to prove first whether it exists within our universe. Otherwise it is fictitious as virtual mathematics is. Therefore, without substantiated proof of the physical evidence, any postulated science proposal even by a world-renowned physicist cannot literally take it as a real science that existed within our universe. For examples, if one of the world greatest astrophysicists tells us that Black Hole provides a physical channel to connecting our universe to another universe, which is absurd. Or there exists a theory of all theories, as a knowledgeable scientist would you take it seriously? If one of the world greatest mathematician found a ten-dimensional subspace within our universe, would you not curious enough to find out whether does it exist within our universe? Nevertheless, we are all humans; that includes all the world eminent scientists, scholars, mathematicians, and philosophers (i.e., past, present and future) are not perfect.

It is well known to us: In mathematics, every postulation needs to prove first that there exists a solution, before searching for the solution. Yet, there is no guarantee that we will find the solution. Nevertheless, in science, it seems to me that we do not have a clear criterion as mathematics does: to prove first that an analytical solution exists within our universe before verifying experimentally whether it exists. Without such a criterion, fictitious, virtual, or even fake sciences emerge, as have been already happening everywhere. This is one of our aims to show that a scientific criterion is needed to prove first that an analytical solution exists within our temporal (i.e., t > 0) universe and then follows up by experimental verification for the existence. In other words, any analytical solution needs to satisfy the basic boundary conditions of our universe: temporal and causal (i.e., t > 0) and dimensional; otherwise it is virtual as mathematics. Fundamental principle of superposition in quantum mechanics is one of the examples; that we have proven it only exists within a timeless virtual subspace and it does not exist within our temporal universe.

Since time and substance coexisted, every subspace takes time to create and the created subspace cannot be used to trade back the time that has been used for the creation. In which we have seen that distance is time and time is distance within our universe. In other words, everything within our universe has a price tag in terms of an amount of energy ΔE and a section of time Δt to create. And this amount of energy and time (i.e., ΔE, Δt) is also a representation for every bit of information needs an amount of energy ΔE and a interval of time Δt to create, to transmit, to store, to compute, and to delete or to destroy, as represented by the Heisenberg uncertainty relationship (i.e., $\Delta E \cdot \Delta t \geq h$, $\Delta v \cdot \Delta t \geq 1$).

As we have already seen in Chapter 4, information can be transmitted either inside or outside a quantum limited subspace (QLS) [14].
For digital-time transmission, we have shown that using a broader
bandwidth (i.e., Δv) carrier has the advantage for narrower time duration (i.e., Δt) for rapid transmission, but wider bandwidth Δv is also
more vulnerable for noise perturbation. On the other hand, for frequency-digital transmission, we prefer a narrower bandwidth instead.
Although it needs a longer duration of Δt for transmission, it has the
advantage of lower noise perturbation.

The inherent advantage of using analog information-transmission has
a higher information capacity than time-digital (or per Δv for frequency-
digital) transmission, but digital signal can be repeated (i.e., refreshed) and
analog cannot. In which we see that primary advantage of using digital
technology in communication is for noise immunity and not just for simplicity in transmission. Nevertheless, there is a price to pay for rapid transmission by using high-speed carrier of electro-magnetic carrier (e.g., light).

In view of either time-digital or frequency-digital transmission, they
are basically using intensity (i.e., ΔE) for transmission that is limited by
the Heisenberg Uncertainty Principle or equivalently communication
outside the QLS domain. Of which we have shown that (i.e., in Chapter
4), complex amplitude information can be exploited within QLS. But
complex amplitude transmission is limited by the mutual coherence
length of the information carrier as given by $d = c/\Delta v$, where Δv is the
bandwidth of the carrier (e.g., quantum state energy bandwidth).

Quantum computing is basically exploiting the simultaneous quantum states for computing; firstly we see that cross-simultaneous states
(e.g., v_1, v_2) cannot be utilized for complex amplitude communication
since they are mutually incoherent. Secondly, if we split a quantum state
radiation into two paths for simultaneous complex communication, it
will be again limited by the quantum state bandwidth Δv_1. Therefore, it
is their mutual coherent length (i.e., $d = c/\Delta v_1$) of the quantum radiator
that limits the information processing capability, since time is distance
and distance is time within our universe. In other words, complex-
amplitude quantum computing is possible within the QLS, but the
computing capability is limited by the quantum state bandwidth Δv_1.

We further note that quantum entanglement is basically communicated within the QLS that is depending on Pauli's half-spin quantum
exclusive principle. In view of the Bohr's atomic model, it is easy to
discern that particles entanglement is relied on quantum state half-spin
of $h\Delta v/2$ which is an electro-magnetic wave, for which particles entanglement cannot beyond the speed of light. Secondly the entangled
distance (i.e. $d = c/2\Delta v$) is limited by $\Delta v/2$ of the Pauli-Heisenberg
uncertainty relationship of Eq. (7.19). In view of mutual entangling
distance of $d = c/(2\Delta v)$, quantum entanglement communication is

operating within the QLS, instead and outside of QLS, since particle quantum radiant frequency v for entanglement is generally higher because of very small atomic size, which has a broader half-spin quantum state bandwidth that limits the entangled distance between particles.

7.5   Look Back at Schrödinger's Quantum Mechanics

In view of the paradox of Schrödinger's cat, it has been puzzling quantum physicists for over eighty five years and has been debating by world's renowned physicists since 1935; and still debating. The overlook of the paradox is very human, since we are imperfect and very limited. As we all agreed that; every law that includes the fundamental law of physics and paradox were made to revise or to be broken. In which we should anticipate the changes and to welcome the fact of science is also temporal (i.e., t > 0). For example, Newtonian space has been treating time as an independent variable for years, but later we found out Newtonian space has violated the fundamental boundary condition of our temporal universe; substance and time coexist. The paradox of the Schrödinger's cat cannot be the exception, where we have found that the fundamental principle of superposition is timeless. Of which the simultaneous exist multi-quantum states of a particle does not exist within our universe. And the reason is very simple. If we look at the Bohr's atomic model shown in Figure 7.4, I assumed that Schrödinger might have used this model to develop his viable Schrödinger equation and his wave equation.

In viewing this fantastic Bohr model, we see that it has treated atom as point-singularity approximation; it has no dimension and no coordinates. One of the most valuable information provided by this model must be the quantum state energy hv such that Schrödinger was capable of developing his sophisticated quantum mechanics. However, as we look at the model from information theory standpoint, no matter how fancy mathematical maneuvering that one uses, any solution generated will be limited by the assumed quantum state hv, which is not a time-limited (i.e., Δt) and band-limited (i.e., Δv) realizable radiation. Although Schrödinger equation has given us for almost a century many useful solutions in quantum physics, it is the limited quantum state information of hv brought us with unthinkable nonexistent solutions, such as the non-existent superposition principle, which created a fantasy time-less quantum world that is fictitious as mathematics.

There is however a fringe-benefit of assuming quantum state energy as a continuous wave radiator (i.e., hv); it simplifies the mathematical evaluation. But the solution as obtained for using Bohr model is the physically reality limited, since quantized energy ΔE should be time limited Δt. In other words, every quantum state radiation has to be time and band limited (i.e., Δt and Δv). In practice, a band and time limited (i.e., Δv, Δt) radiation (or wavelet) is a physical realizable assumption, and it should have had used when Schrödinger developed his quantum mechanics. This might have been one of the several reasons that our predecessors have had overlooked this issue and have had treated quantum state energy as a "continuous with no-bandwidth" electro-magnetic wave emitter.

Now let me provide a dual-quantum states (i.e., a multi-quant states scenario) atomic model as an example depicted in Figure 7.6, of which two separated quantized wavelet radiators are assumed existed within the atomic model. This model shows us that the dual-quantized wavelets will very "unlikely" occur simultaneously, as in contrast with the superposition principle of continuous wave radiators as depicted in Figure 7.5. Thus we see that, if the fundamental principle of superposition was developed using the wavelets assumption (i.e., Δt and ΔE)

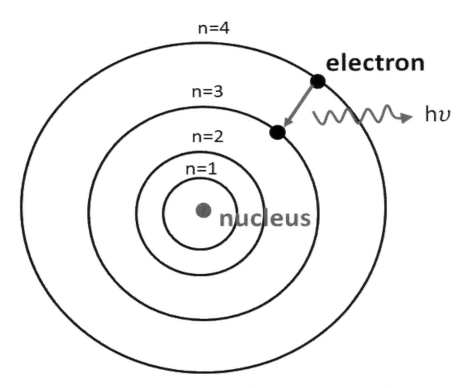

Figure 7.5 A Bohr atomic model; h is the Planck's constant and v is the quantum state radiation frequency v.

instead of continuous wave model, the simultaneously existed quantum states would not have developed as a fundamental principle in quantum mechanics. Then we won't have had the legacy of the paradox of the Schrödinger's cat and the fantasy of the timeless quantum wonderland.

Advances of science are temporal (i.e., t > 0) and new ideas and discoveries as well as the replacement and improvement are the inevitable events in every sector of science. And the atomic model is proposed by Bohr since 1911, which is over a century ago, may be it is time to be revised for better solution, as we move on to the sub-atomic region. In order to avoid fictitious solution that emerges, a more reliable temporal atomic model as given by Figure 7.7 can be used at this time. In which we see that a time- and band-limited quantized (i.e., $\Delta E = h\Delta v$) state energy is schematically shown within the model.

In which we show a modified Bohr atomic model is embedded within a temporal (i.e., t > 0) subspace and the quantum state energy is

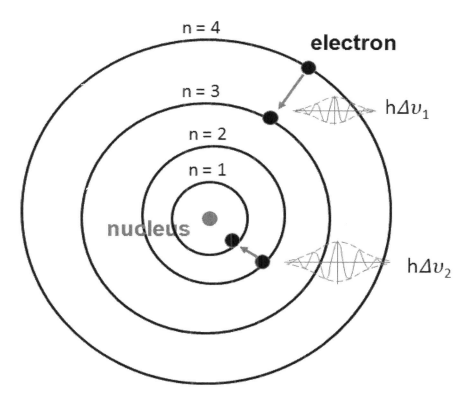

Figure 7.6   A dual-quantum states atomic model; h is the Planck's constant and Δv is the bandwidth quantum state radiation frequency v.

represented by a wavelet of a band- and time-limited radian energy (i.e., $\Delta E = h\,\Delta v$), instead of a continuous wave emitter. This model will provide a step closer to a more realistic pocket wavelet edition in developing a "temporal quantum machine," although the derivation of quantum mechanical equation (i.e., similar to the Schrödinger equation) can still be used a continuous wave emitter (i.e., hv) for mathematical simplicity; after all science is a law of approximation. In view of this new model of Figure 7.7, we would anticipate the wave equation as derived from this new edition will be in the form approximately as given by,

$$\psi(x; \omega t) = A \exp[-\alpha(t - t_0)^2]exp\ (i\ kx - \omega t), t > 0 \qquad (7.20)$$

where A is an arbitrary constant, t_0 is an arbitrary time delay, $\omega = 2\pi v$, v is the quantum state frequency, x is a spatial variable, α is an arbitrary constant, and t is a forward variable as specified $t > 0$. For which the eigen-value of the wavelet can be shown as given by,

$$\omega(t - t_0) = A \exp[(t - t_0)]cos(\omega t), t > 0 \qquad (7.21)$$

In which we see that the quantum state wavelet is limited by Δt shown in Figure 7.8, instead of presenting as a time-unlimited continuous wave radiator.

Needless to say that, as for a multi-quantum states wavelet representation, we would anticipate that the quantized wavelets will be in forms as given by,

$$\psi\{\omega_n[(t - t_0) - \Delta t_n]\} = A\ \exp\left\{-\alpha[(t - t_0) - \Delta t_n]^2\right\}$$
$$\cos\{\omega_n[(t - \Delta t_n]\},\ t > 0; \text{forn} = 1,\ 2,\ \ldots N \qquad (7.22)$$

where $\omega_n = 2\pi v_n$, Δt_n is the time separation between n = 1 and nth quantum state. Figure 7.9 shows a scenario for multi-quantum states emission from a singularity atom where we see that the quantum state energy emits not simultaneously. If Schrödinger has had considered the time- and band-limited issue of the quantum states, he might not have had developed his fundamental principle of superposition, which unfortunately does not exist within our temporal universe.

Let us look back at an example as depicted in Figure 7.10, in which we have two time-limited quantum state wavelets that are plunging into cascaded timeless-temporal subspaces. Figure 7.10(c) shows the corresponding response within a timeless space in which the two time-limited wavelets collapsed at t = 0; we see that input wavelets lost their original

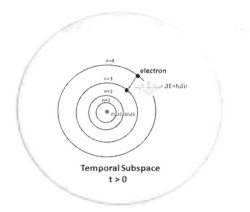

Figure 7.7 A temporal (t > 0) atomic model: $\Delta E = h\Delta v$ represents a time- and band-limited quantized state energy (i.e., one quantum of energy).

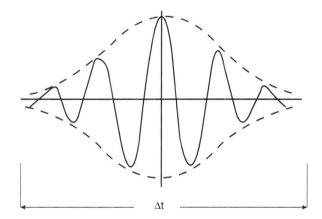

Figure 7.8   A time-limited quantized state wavelet (i.e., $\Delta E = h\Delta v$).

temporal personalities. If we plunge this timeless response of Figure 7.10(c) into a temporal (i.e., t > 0) space of Figure 7.10(d), then the response within a temporal space can be seen in Figure 7.10(e). We see that the response complies with the temporal and causality (i.e., t > 0) constraints; for which the response shows no sign of preserving the original wavelet properties; temporally and spatially. Once again we have proved that the superposition principle does not exist with a cascaded (i.e., timeless-temporal) system as

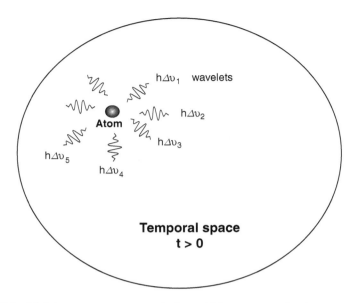

Figure 7.9 Multi-quantum states emission within a temporal subspace.

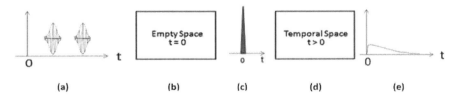

Figure 7.10 A system representation for a cascaded timeless to temporal space representation. (a) A set of time limited wavelets plunge into a timeless space of (b). (c) The output from the empty space. (d) The corresponding response within a temporal space. In which we see that output loses all the original input personality as shown in (e).

shown in Figure 7.10. In which we have again shown that superposition is timeless and only exists in an empty space as mathematics does.

Therefore, we stress that it is a "serious mistake" to implant a timeless supposition principle (i.e., an analytical solution) within our temporal universe, even though by ignoring the physical realizable issue. Beside the timeless and temporal spaces cannot coexist, the fact is that quantum mechanist could do it, since quantum mechanics is mathematics. But we have shown that the overall response within our temporal universe has lost all its temporal and spatial personalities.

It is however a "more serious mistake" if one is asking a timeless superposition to behaves "timelessly" (i.e., $t = 0$) within our temporal (i.e., $t > 0$) universe. This is precisely the reason why some quantum physicists claimed that within particle size regime; particles behave like "Alice in the Wonderland," but they have forgotten time is distance and distance is time within our universe. For example, no matter how small the particle is or the separations are, their sizes and separations cannot be ignored; since $d = c \cdot \Delta t$, where Δt is the time separation between quantum state wavelets and c is the speed of light (i.e., a huge quantity).

This is the moment for us to look back at quantum mechanics which was developed over 85 years ago. After we have learned the "timeless" fundamental principle (i.e., an analytical solution) as mathematics is, we may understand quantum mechanics actually is. As quoted from one of the many Richard Feynman's remarks, "If you think you understand **quantum mechanics**, you don't understand **quantum mechanics**." Now you might learn the part that you do not understand; quantum mechanics is a timeless machine as mathematics does, and her fundamental principle and her quantum world are timeless. Nonetheless, in principle, we can build a temporal (i.e., $t > 0$) quantum machine, provided the quantum machine is built on a temporal (i.e., $t > 0$) subspace platform as we have shown in a preceding section 7.2.

Furthermore, practically all the fundamental laws of science are point-singularity assumption, and it would be very difficult to describe a multi-dimensional physical problem, such as our temporal universe. In which I have shown our universe was created from a dimensionless law of energy (i.e., $E = mc^2$) to a three-dimensional time-dependent space. As I used subspace (i.e., set theory) representation, it reduces the complexity of mathematical manipulation which provides us with simpler description, as in contrast if one uses point-singularly approach. Furthermore, as from our preceding presentation, we found that science (i.e., analytical solution) is mathematic, but mathematics is not "necessary" equal to science, unless her analytical solution satisfies the boundary condition of our universe: dimensionality, temporal, and causality condition (i.e., $t > 0$).

7.6   Remarks

Since Schrödinger's quantum mechanics was developed on the top of timeless subspace platform, his machine is timeless. His fundamental principle of superposition is also timeless which does not exist in our temporal universe. We have attempted to develop a time-dependent or temporal quantum mechanics that exists within our temporal universe. We have shown that solution as obtained from an atomic model which is submerged within a temporal subspace guarantees the solution existed within our universe. Otherwise, fictitious virtual solution emerges which produces unthinkable consequences. For example, the paradox of Schrödinger's cat has been persisted over eight decades of debates since the disclosure in 1935. In which we also see that the whole Schrödinger's quantum world is timeless that includes the core of his fundamental principle of superposition principle is timeless. But timeless superposition does not exist within our universe. As a matter of fact, those "simultaneous existent" quantum states for quantum computing and the "instantaneous" particle entanglement for communication only existed in a virtual timeless mathematical space. And we have also shown that, in principle, a temporal (i.e., t > 0) quantum mechanics can be built if the atomic model is anchored within a temporal platform of which the solution as obtained by temporal quantum machine will comply within the boundary condition of our temporal universe; causality (i.e., t > 0) and dimensionality. Nevertheless, Schrödinger quantum mechanics has produced unaccountable excellent results for practical application, as long its solution does not violate the causality (i.e., t > 0) condition of our temporal universe of which his fundamental principle of superposition (i.e., t = 0) has violated.

We stress that it is a serious mistake if one forces a timeless superposition principle (i.e., t = 0) into our temporal universe (i.e., t > 0) and anticipated the fundamental principle behaves timelessly within our temporal universe. And this is precisely that we need a criterion for any newly discovered mathematical-science; firstly is to prove that it exists within the boundary condition of our temporal universe—causality (i.e., t > 0) and dimensionality—and then verifies by repeated experiments.

Finally, a countered example of "Schrödinger's cat" is that we have a set of observers who take turns to look into the Schrödinger's box. If one observer opens up the box after looks into it, and then closes the box for the next one to observe. My question is that how many times the cat has to suffer for dying? If Schrödinger has had thought through the consequences of his cat, he might not have had hypothesized his half-life cat!

In addition: very likely this would be my last book before we move to an independent living facility nearby. Yet, I have a viable advise to all the scientists and researchers: It is not how rigorous the mathematics is; it is the essence of a physical realizable paradigm. For instance; if one uses a nonphysical realizable model, unthinkable solution may emerge such as the timeless superposition principle. Mathematics evaluation and computer animation are virtual, one should not treated their solution and simulation are real, unless the models they used are physical realizable.

References

1. E. Schrödinger, "Die gegenwärtige Situation in der Quantenmechanik (the Present Situation in Quantum Mechanics)," *Naturwissenschaften*, vol. 23, no. 48, 807–812 (1935).

2. F. T. S. Yu, "Time: The Enigma of Space," *Asian J. Phys.*, vol. 26, no. 3, 149–158 (2017).

3. F. T. S. Yu, *Entropy and Information Optics: Connecting Information and Time*, 2nd ed., CRC Press, Boca Raton, FL, 2017, 171–176.

4. L. Susskind and A. Friedman, *Quantum Mechanics*, Basic Books, New York, 2014, 119.

5. N. Bohr, "On the Constitution of Atoms and Molecules," *Philos. Mag.*, vol. 26, no. 1, 1–23 (1913).

6. K. Życzkowski, P. Horodecki, M. Horodecki, and R. Horodecki, "Dynamics of Quantum Entanglement," *Phys. Rev. A*, vol. 65, 012101 (2001).

7. T. D. Ladd, F. Jelezko, R. Laflamme, C. Nakamura, C. Monroe, and L. L. O'Brien, "Quantum Computers," *Nature*, vol. 464, 45–53 (March 2010).

8. E. Schrödinger, "Probability Relations between Separated Systems," *Mathematical Proc. Cambridge Philos. Soc.*, vol. 32, no. 3, 446–452 (1936).

9. L. D. Landau and E. M. Lifshitz, *Quantum Mechanics*, Pergamon Press, Oxford, 1958, pp. 50–128.

10. W. Heisenberg, "Über den anschaulichen Inhalt der quantentheoretischen Kinematik und Mechanik," *Zeitschrift für Physik*, vol. 43, 172 (1927).

11. W. Pauli, "Über den Zusammenhang des Abschlusses der Elektronengruppen im Atom mit der Komplexstruktur der Spektren," *Zeitschrift für Physik*, vol. 31, 765 (1925).

12. F. T. S. Yu, *Introduction to Diffraction, Information Processing and Holography*, MIT Press, Cambridge, MA, 1973, p. 94.

13. F. T. S. Yu, "The Fate of Schrodinger's Cat," *Asian J. Phys.*, vol. 28, no. 1, 63–70 (2019).

14. F. T. S. Yu, "Information-Transmission with Quantum Limited Subspace," *Asian J. Phys.*, vol. 27, no. 1, 1–12 (2018).

Appendix

Aspects of Particle and Wave Dynamics

In view of temporal (t > 0) universe, every subspace within our universe is a time-dependent subspace, in which time is coexisted with every subspace and time is a "dependent forward variable." The speed of time within our universe is dictated by the speed of light that our universe was created with. In other words, any subspace (i.e., substance) within our universe is a time-depending or temporal (t > 0) subspace in which the time speed within the subspace is unison (or in pace) with the time speed of the entire universe; otherwise the subspace cannot be existed within our universe. For example, the time speed within a subspace at the edge of our universe is at the same pace as the subspace closer to the center of our universe. On the other hand, if the time speed within a subspace runs faster or slower than the time speed of our universe, then the subspace cannot be existed within our universe. An extreme example is that a timeless (i.e., empty) space is a virtual mathematical space; firstly it is not a temporal space and secondly the space has no time for which it cannot exist within our temporal (t > 0) universe.

Since every subspace within our universe was created by an amount of energy ΔE and a section of time Δt that creates the subspace, it cannot bring back the section of time Δt that had been used for the creation. There is, however, a profound relationship between energy and mass (i.e., ΔE, Δt) by virtue of Einstein's energy equation (i.e., $E = mc^2$). Of which there exists a duality between energy ΔE and bandwidth Δv. Nevertheless, without the coexistence with time, the duality collapses at timeless (t = 0) empty space, which is not a subspace within our universe.

Let us look at energy ΔE and bandwidth Δv duality; firstly it must be the existence of waves, otherwise it will not have the aspect of bandwidth. So what is wave? Wave can only exist within a temporal (t > 0) subspace, since physical substance and time are coexisted; otherwise, waves cannot propagate with time. In other words, waves can only exist within a temporal medium (i.e., substance) that creates wave. This is also one of the key factors that many quantum physicists believe that quantum leaps radiation propagates within an empty space. Aside the non-physical realizable issues, substance (i.e., radiation) and empty space are mutually exclusive; the fact is that wave cannot exist within an empty (i.e., timeless) space. Secondly, every physical radiation has to be band and time limited (i.e., Δv, Δt,), as given by the Heisenberg Uncertainty Principle,

$$\Delta v \cdot \Delta t \geq 1$$

In which we see that there exists an energy and bandwidth relationship as given by,

$$\Delta E = h\Delta v$$

where h is the Planck constant. This is the well-known quantum leap energy or a "quanta" of electro-magnetic radiation, in which we see a time-limited package of energy ΔE that travels at the speed of light within our temporal universe.

Nevertheless, root of particle-wave duality was originated by the acoustic wave dynamics where acoustic wave travels "longitudinally." For example, an acoustic wave generated by a Guitar is produced by a vibrating string that satisfies the wavelength selectivity boundary condition that causes a pulse of wavelet (i.e., time and band limited) travels within our atmospheric space. Since it is a longitudinal wave, an acoustic wave depends on media such as solid material, liquid, or air for transmission. In other words without such media, any acoustic wave cannot be created within the medium. Thus we see that the particle-wave dynamics is a mathematical description; by using the dynamics wave propagation to predict the behavior of an assumed particle agitation. Notice that our illustration is in fact a fix-ended string vibration instead of a particle; in which we see that it is actually a string to wave dynamics.

On the other hand, electro-magnetic wave is a transversal wave which propagates within electro-magnetic (EM) medium (i.e., permeability ε and permittivity μ medium). Again we note that empty space has no medium within it; an electro-magnetic wave cannot exist within an empty space. In view of particle-wave duality description, electro wave dynamics exists within our temporal (t > 0) space and it is not the implication that an EM wave was created by means of a physical particle (i.e., photon) agitation within the space, since every physical particle requires a rest mass. Notion of photon behaves as particle has been well accepted, but it is difficult to reconcile with the relativity theory for the assumption; photonic particle has an empty mass, for which I would regard every photon as a virtual particle which is attached with a quanta energy $h\Delta v$. Although mass and energy are equivalent as from relativity theory (i.e., Einstein's energy equation) stand point, but a quantum of electro-magnetic energy is not created by the annihilation of a rest mass, instead it is in form of EM energy releases by each quantum state leap radiation hΔv. It is for this reason that packet of energy E to wavelet Δv duality is a better presentation as to describe the dynamics of quantum state behavior. Since energy has different forms such as, potential, kinetic, chemical, radiation, nuclear, and others, photonic energy is a packet wavelet radiation derived from a quantum leap of an electron within an atom. Particle-wave duality is describing a packet of energy ΔE to wavelet (i.e., Δv) dynamics or simply (ΔE, Δv) duality. The energy transfer from a higher quantum state to a lower quantum state releases energy in EM wave, very similar to a fix-end string oscillation from a Guitar. Therefore, it is more suitable to use energy to wavelet (ΔE, Δv) duality to describe the quantum leap wave dynamics, otherwise the particle-wave duality gives us a notion that photon is a physical particle instead of a quantum of energy (i.e., quanta). In which we see that a time limited wavelet represents a package of energy and a package of energy is a time limited wavelet.

Index